OPCS Surveys of Psychiatric Morbidity in Great Britain

Report 8

Adults with a psychotic disorder living in the community

Kate Foster

Howard Meltzer

Baljit Gill

Kerstin Hinds

London: HMSO

ISBN 0 11 691670 2

Notice

On 1 April 1996 the Office of Population Censuses and Surveys and the Central Statistical Office merged to form the Office for National Statistics. The logo of the new Office appears on the front cover of this report but the report series name - 'OPCS Surveys of Psychiatric Morbidity in Great Britain' - remains unchanged, to preserve the continuity of the 8-report series. Reference to OPCS inside the report are also unchanged since it was already in production at the time of the merger.

HMSO

Published by HMSO and available from:

HMSO Publications Centre
(Mail, fax and telephone orders only)
PO Box 276, London SW8 5DT
Telephone orders 0171 873 9090
General enquiries 0171 873 0011
(queuing system in operation for both numbers)
Fax orders 0171 873 8200

HMSO Bookshops
49 High Holborn, London WC1V 6HB
(counter service only)
0171 873 0011 Fax 0171 831 1326
68–69 Bull Street, Birmingham B4 6AD
0121 236 9696 Fax 0121 236 9699
33 Wine Street, Bristol BS1 2BQ
0117 9264306 Fax 0117 9294515
9–21 Princess Street, Manchester M60 8AS
0161 834 7201 Fax 0161 833 0634
16 Arthur Street, Belfast BT1 4GD
01232 238451 Fax 01232 235401
71 Lothian Road, Edinburgh EH3 9AZ
0131 228 4181 Fax 0131 229 2734
The HMSO Oriel Bookshop
The Friary, Cardiff CF1 4AA
01222 395548 Fax 01222 384347

HMSO's Accredited Agents
(see Yellow Pages)

and through good booksellers

Authors' acknowledgements

We would like to thank everybody who contributed to the survey and the production of this report. We were supported by our specialist colleagues in OPCS who carried out the sampling, fieldwork, coding and editing stages.

Great thanks are due to the interviewers who worked on the survey and to the psychiatrists who conducted the SCAN interviews.

The project was steered by a group comprising the following, to whom thanks are due for assistance and specialist advice at various stages of the surveys:

Department of Health:
Dr Rachel Jenkins (chair)
Dr Terry Brugha
Ms Antonia Roberts
Mr Alan Madge

Psychiatric epidemiologists:
Professor Paul Bebbington
Professor Glyn Lewis
Dr Mike Farrell
Dr Jacquie de Alarcon

Office of Population Censuses and Surveys:
Ms Jil Matheson
Dr Howard Meltzer
Ms Baljit Gill
Ms Kerstin Hinds

Most importantly, we would like to thank all of the participants in the survey for their time and co-operation.

Contents

Notes

1 Tables showing percentages

The row or column percentages may add to 99% or 101% because of rounding.

The varying positions of the percentage signs and bases in the tables denote the presentation of different types of information. Where there is a percentage sign at the head of a column and the base at the foot, the whole distribution is presented and the individual percentages add to between 99% and 101%. Where there is no percentage sign in the table and a note above the figures, the figures refer to the percentage of people who had the attribute being discussed, and the complementary percentage (who did not have the attribute) is not shown in the table.

The following conventions have been used within tables showing percentages:

- \- no cases
- 0 values less than 0.5%

Small bases
Very small bases have been avoided wherever possible because of the relatively high sampling errors that attach to small numbers. Where small bases cannot be avoided, percentages have been shown but those based on fewer than 40 cases are shown in brackets.

2 Tables showing results of logistic regression analysis

(see also Section 1.5 and Appendix B)

The tables list only the variables which were included in the final logistic regression model.

For each independent variable included in the model, one category was defined as the reference category. This is indicated by an odds ratio (OR) of 1.0.

The OR for other categories indicate the factor by which the odds for that category of informants differed from the odds for the reference category.

Asterisks indicate which odds ratios are significantly different ($p<0.05$) from 1.0. Significance tests on these unweighted data should be treated with some caution and used only as an indication of the reliability of results.

Summary of main findings

Background

This is the eighth and final report on the OPCS surveys of psychiatric morbidity in Great Britain. Four separate surveys were commissioned by the Department of Health, the Scottish Home and Health Department and the Welsh Office and carried out between April 1993 and August 1994. These covered adults aged 16 to 64 years living either in private households (two separate surveys) or in institutions or who were identified as homeless.

The analysis in this report brings together people with a psychotic disorder from the first three surveys. Adults were included in the analysis if they were deemed to be living in private households.[1] This definition included people in premises originally approached as part of the institutional sample and who lived, for example, in group homes or recognised lodgings: these are referred to as people living in supported accommodation.

The main focus of this report is to investigate the behaviour and circumstances of adults with psychosis living in households (rather than in hospitals or hostels) and to identify characteristics associated with different aspects of their functioning and circumstances. This is achieved by multi-variate analysis for the whole sample and by the inclusion of a small number of case studies.

The sample used for the analysis in this report is not representative of all people with psychotic disorders because individuals in the different samples had different probabilities of selection and the results have not been re-weighted to compensate for this (see Section 1.5). Thus percentages shown in tables indicate prevalence for this sample and cannot be regarded as estimates for all people with psychosis living in residential households. The main focus of the results is therefore on factors associated with the behaviour and circumstances of people with psychosis.

Medication (Chapter 2)

- The probability of being on anti-psychotic or antidepressant medication was associated with age, and was highest for informants aged 45-54. *(Table 2.2)*
- Of those on anti-psychotic medication, about three in ten were receiving depot injections. People classified to manual rather than non-manual social class groups had a higher probability of being on this type of medication. *(Table 2.4)*
- Non-compliance with dosage for medication was more likely among sample members with a high CIS-R score, which is a measure of neurotic symptoms. *(Table 2.7)*
- The odds of having stopped or refused treatment for a psychotic condition were greater for those with the highest level of educational qualifications, A level or above. *(Table 2.9)*

Use of services (Chapter 3)

- Three fifths of this sample of people with psychosis had consulted a GP in the past year about a mental, nervous or emotional problem. Informants living in supported accommodation (group homes or recognised lodging) were less likely than those living in private households to have done so. *(Table 3.2)*

- One fifth of the sample had been an in-patient in the previous year for a mental problem. The odds were higher for younger informants, men and people living in private households rather than supported accommodation. *(Table 3.4)*

- One half of the sample had been an out-patient during the previous year for a mental, nervous or emotional problem. Again, people in private households, and especially those living with other people, were more likely than those in supported accommodation to have done so. *(Table 3.5)*

- Almost one half of the sample had received a domiciliary visit from a community psychiatric nurse, one quarter had been visited by a social worker and one in six received visits from a home care worker. The factors associated with receiving visits varied according to the type of visit. Informants living in supported accommodation were more likely than those in private households to have been visited by a social worker or home care worker. *(Table 3.7)*

- Overall, the characteristic most strongly associated with having used at least one of the four types of service considered was the CIS-R score. Informants with a score of 18 or more, indicating a higher level of neurotic symptoms, were most likely to have used one of the services. The odds were also higher for informants who were classified as White and those who did not have a longstanding physical illness. *(Table 3.9)*

- Informants with a high CIS-R score (18 or over) were also most likely to have decided not to see a doctor when others thought that they should. *(Table 3.11)*

Activities of daily living (ADL) (Chapter 4)

- More than three fifths of the sample reported difficulties with at least one of the activities of daily living (ADL) covered by the surveys. People were least likely to have difficulties with personal or medical care and most likely to have difficulties with dealing with paperwork. *(Table 4.1)*

- Either physical illness or the degree of neurotic symptoms were strongly associated with most ADL difficulties. Neither was associated with difficulties with paperwork: men and informants with no educational qualifications were most likely to have problems with these. Informants living in supported accommodation were more likely than those in private households to have difficulties with medical care and household activities. *(Tables 4.2 and 4.3)*

- The probability of having four or more ADL difficulties was associated with having a physical illness, a CIS-R score of 18 or more and living in rented or supported accommodation. *(Table 4.5)*

- At least two thirds of informants who had difficulties with ADL said that they needed help and most of these received it. The probability of receiving help from a health professional or voluntary worker was higher for informants living in supported accommodation. *(Table 4.7)*

Economic activity and finances
(Chapter 5)

- About one half of this sample of people with psychosis were classed as unable to work, one in five was in employment and one in eight was unemployed. *(Table 5.1)*

- A large number of characteristics were associated with being unable to work. People with a CIS-R score above the threshold level of 12, informants aged 35 or over and those previously in manual occupation groups had higher odds of being classified as unable to work. Informants living in supported or rented accommodation were also more likely to be permanently sick, whereas people who were owner occupiers had the highest odds of being in employment. *(Table 5.3)*

- Nine tenths of the sample controlled their own finances and about one half of those who did received a state benefit relating to invalidity or disability. The probability of receiving these benefits was higher for men and for people who also had a longstanding physical illness. *(Table 5.5)*

Social functioning (Chapter 6)

- About one quarter of this sample were identified as having a small primary support group. People living in supported accommodation or in private households but not with family members were more likely than those living with members of their family to have a small support group. *(Table 6.2)*

- One third of the sample had a severe lack of perceived social support. Again the probability of being in this category was higher for people not living with members of their family and for those with a CIS-R score of 18 or over. In addition, informants with a small primary support group had much greater odds of perceiving a severe lack of social support. *(Table 6.4)*

- The factors associated with taking part in few leisure activities differed from those associated with social support. People with no educational qualifications and men were more likely than those with qualifications and women to take part in fewer than four leisure activities. *(Table 6.6)*

- People living in supported accommodation were more likely than those living in private households to visit either a day centre, a club for people with disabilities or a social club. Adults with the highest qualifications, A-level or above, were most likely to be attending an Adult Education or Adult Training Centre. *(Table 6.7)*

Tobacco, alcohol and drugs (Chapter 7)

- Around three fifths of this sample were regular smokers and two fifths smoked 20 or more cigarettes a day. People in supported or rented accommodation were more likely than those who owned their accommodation to be both regular and heavy smokers. *(Table 7.2)*

- A smaller proportion of this sample were heavy drinkers than in the general population in private households. As in the general population, men and younger people were more likely than women and older groups to be regular drinkers. The only characteristic associated with drinking more than the recommended sensible level was sex: again, men had higher odds. *(Table 7.4)*

- The most common categories of extra-medical or illicit drugs used by this sample were cannabis and hypnotics and the factors associated with using drugs varied for the two types. Informants with a CIS-R score of 18 or more were most likely to be taking hypnotics whereas use of cannabis varied significantly with the age of the informant, being most likely among the 16-34 age group. *(Table 7.6)*

The circumstances of people with psychosis (Chapter 8)

- Many of the measures used as dependent variables throughout this report were associated with each other. There was, for example, a strong association between being on antipsychotic or antidepressant medication and having used some health services in the past year. The variables which showed least inter-relationship with others were alcohol consumption and use of illegal drugs.

- There was a strong association between having four or more ADL difficulties and being classed as unable to work. The odds of having both of these characteristics were higher among people living in supported or rented accommodation, and increased with CIS-R score. *(Tables 8.1 and 8.2)*

- Having a small primary support group and a severe lack of perceived social support were also closely associated. Again, people living in supported accommodation and those with a high CIS-R score, of 18 or more, had a higher probability of falling into this category. *(Table 8.3)*

- There were weaker associations between the different health-related behaviours considered in Chapter 7 - smoking, drinking and use of drugs. No characteristics were associated with being a heavy smoker and drinker and having used illegal drugs. *(Table 8.1)*

Notes and references

1 This was coded by interviewers who are familiar with the usual definition of a household. The definition is: one person or a group of people who have the accommodation as their only or main residence and (for a group of people) either share at least one meal a day or share the living accommodation.

1 Background

1.1 Focus of the report

This report focuses on adults aged 16 to 64 with a psychotic disorder who were interviewed for the OPCS surveys of psychiatric morbidity. The informants included in the analysis for this report were all living in household units and were identified either from the surveys of people in private households or the survey of institutions. The standard survey definition of a household is:

> one person or a group of people who have the accommodation as their *only* or *main* residence, and (for a group of people) *either* share at least one meal a day *or* share the living accommodation.

Earlier reports in this series have presented estimates of prevalence of psychosis for the population in private households and institutions and for homeless people.[1-3] This report has a different focus which is to investigate the behaviour and circumstances of adults with psychosis living in households, rather than in large institutions, and to identify characteristics associated with different aspects of their functioning and circumstances. This is achieved by multi-variate analysis for the whole sample and by the inclusion of a small number of case studies.

The report considers:

- use of medication including whether anti-psychotic medication is administered by injection or orally, non-compliance with medicinal regimes and refusal of treatment
- use of services including consultations with GPs, in-patient stays, out-patient visits and domiciliary visits
- difficulties with activities of daily living and who provides help with these activities
- social functioning including size of social networks, perceived social support and use of leisure time
- economic activity and income, including whether in receipt of benefits
- health-related behaviours including smoking, drinking and drug use
- the inter-relationship between different behaviours and circumstances covered in the analysis

1.2 The surveys of psychiatric morbidity

The OPCS Surveys of psychiatric morbidity in Great Britain were commissioned by the Department of Health, the Scottish Home and Health Department and the Welsh Office. They aim to provide up-to-date information about the prevalence of psychiatric problems among adults in Great Britain as well as their associated social disabilities and use of services.

Four separate surveys were carried out from April 1993 to August 1994:

i) 10,108 adults aged 16 to 64 years living in private households (fieldwork: April 1993 - September 1993)

ii) a supplementary sample of approximately 300 people aged 16 to 64 years with psychosis living in private households (fieldwork: October 1993 - December 1993)

iii) 1,191 people aged 16 to 64 years living in institutions specifically catering for people with mental illness (fieldwork: April 1994 - July 1994)

iv) 1,166 homeless people aged 16 to 64 years living in hostels for the homeless or other such institutions. This sample also included people sleeping rough (fieldwork July 1994 - August 1994)

1.3 The sample used in this report

The analysis in this report brings together people with a psychotic disorder from the first three of the surveys of psychiatric morbidity listed above. Adults identified in these surveys were included in the analysis if they were deemed to be living in private households.[4]

The 470 people covered in this report comprise:

- 54 adults with psychosis identified from the private household survey.

- 244 adults from the supplementary sample of people with psychosis living in private households. Some adults originally chosen for this supplementary sample have been excluded from the analysis for this report because they were subsequently found not to meet any of the survey criteria to identify people with psychosis.

- 172 residents with a psychotic disorder identified in the institutional sample who would be regarded as living in residential households, i.e. those living in supported accommodation such as recognised lodgings or small group homes.

Adults with psychosis identified by the survey of homeless people were not included in the sample for this report because only a very small number were considered to be living in private households, for example those living temporarily in private sector leased accommodation. It was probable that this group would have different circumstances to informants on the other surveys but the small number prevented them from being separately identified in the analysis.

1.4 Identification of psychotic psychopathology

Different research strategies were used to identify people with psychosis in each of the three surveys of psychiatric morbidity from which this sample was drawn. The choice of strategy was dependent on several factors: whether the subjects themselves could be interviewed; whether the subject was likely to have knowledge of their condition; whether reliable information could be collected on names of drugs, and whether clinicians were used to carry out SCAN interviews.[5]

Private household survey

The initial stage in identifying people with a psychotic illness living in private households was to use lay interviewers to sift for the possibility of psychotic disorder. The sifting process required informants to score positively on any one of five criteria:

- Subject reported having a psychotic illness

- Subject reported that their GP or other health professional had diagnosed them as having a psychotic illness

- When asked to state their prescribed medication, subjects mentioned drugs which are used to treat psychosis and related conditions

- Injections which subjects said they were receiving were normally given for psychotic conditions

- Subject screened positive on the Psychosis Screening Questionnaire which looked for the occurrence of psychosis-related behaviour: mania, thought disturbance, paranoia, delusions and auditory hallucinations, in the past year[6]

The reference list of drugs and conditions and the Psychosis Screening Questionnaires are given in Appendix A.2.

Clinicians who followed up potential cases were trained to carry out their interviews using SCAN (Schedules for Clinical Assessment of Neuropsychiatry). SCAN is a set of instruments aimed at assessing, measuring and classifying the psychopathology and behaviour associated with the major psychiatric disorders of adult life. Of its four main components, the PSE10 was the one deemed most applicable for the purposes of the survey.[7] The PSE10 itself has two parts. Part One covers, inter alia, anxiety, depressive and bipolar disorders. Part Two includes the psychotic disorders of interest to the survey - schizophrenia, delusional and schizoaffective disorders.

Where it was not possible to carry out a SCAN interview, because of refusals or non-contacts, the OPCS sift data were re-examined taking into account the analysis of the relationship between sift and assessment data for the successful interviews. Those who were on anti-psychotic medication (orally or by injection) and who reported, either directly or indirectly (via doctor's diagnosis), having a psychotic illness were regarded as having a functional psychosis.

Supplementary survey

The same sifting process and clinical assessment procedures were used in the supplementary survey as for the private household survey. The only difference was in the attribution of a diagnosis to those who, for whatever reason, did not have a follow-up SCAN interview.

Subjects for the supplementary survey were obtained from GPs or Mental Health Teams (following Local Research Ethics Committee approval). Doctors and other health professionals were asked to approach their clients who were suffering from a psychotic disorder and ask them to participate in the OPCS survey. Therefore, if the subjects screened positive on any of the five sift criteria, they were confirmed as having a psychotic disorder. There were a few adults chosen for this supplementary sample who said they had no mental health problems, were not taking anti-psychotic medication, and did not have delusions or hallucinations etc. They were excluded from the analysis for this report.

Institutions survey

The process of identifying residents suffering from psychotic disorders in institutions involved:

- asking residents directly what was the matter with them
- asking staff, what was the matter with the subject, (if the subject could not answer but gave informed consent for another person to do so)
- asking residents or carers whether subjects were taking anti-psychotic drugs or having anti-psychotic injections
- establishing whether those living in residential accommodation had contact with any health care professional for a mental, nervous or emotional problem which had been described as a psychotic illness.

Reports of psychosis by subjects or staff, or a positive response to the medication criteria were regarded as indicative of a psychotic illness for the purpose of the analysis used to produce the results in this report. Clinicians were not asked to carry out SCAN interviews in institutions.

1.5 Analysis methods

The sample used for the analyses in this report is, as described above, drawn from three separate surveys. Each of these had a complex sample design and different parts of the sample had different probabilities of selection. Hence results need to be re-weighted in order to produce estimates for the population under study in each of the separate surveys. Combining the samples results in even greater complexity and there is also the likelihood of some duplication of addresses between the various sampling frames used.

In consequence, this combined sample has not been re-weighted to adjust for different probabilities of selection and is not representative of all adults with psychosis living in private households. Hence any percentages presented in this report are indicators of prevalence only within this particular sample and cannot be used as estimates of prevalence among the total population of adults with a psychotic disorder.

Nonetheless, the sample is useful in allowing us to test whether different characteristics are associated with different behaviours or circumstances of people with psychosis. In particular, although all of the people included in the analysis in this report are classified as living in private households, it is probable that there are considerable differences between those identified through the sample of institutions and those selected in the private household samples. Possible differences in service use and in other circumstances of these two groups can be investigated by means of multi-variate analysis on the unweighted data-set. This can be used to indicate which groups have a high or low probability of the event being investigated: as with percentages, odds ratios from the models may not be used as estimates of the odds for the total population.

Logistic regression analysis

The main method of analysis used throughout this report is multiple logistic regression. This method identifies which of a set of independent variables, or characteristics of people in the sample, are associated with a dichotomous dependent variable. The dependent variables were set up to indicate the presence or absence of a particular behaviour or state, for example, informant was on prescribed medication versus not on medication, or informant was a heavy smoker versus not a heavy smoker. The analysis identifies which of the independent variables included in the analysis are most strongly associated with the dependent measure after controlling for the effect of other variables in the model. A forward stepwise method of analysis was used.

The tables showing the results of the logistic regression analyses list only those characteristics which were significantly associated with the dependent measure, although all of the variables described in Section 1.6 were initially included in the analyses. An example of the results is given in Table 3.2 (page 19). In this case the dependent variable is whether the informant had consulted a GP in the past year for a mental condition. Two variables - living arrangements and CIS-R score - were included in the final model. Thus each of these variables was, after allowing for the effects of the other, significantly associated with the probability of having consulted a GP.

For each variable included in the model, logistic regression produces an estimate of the odds of the event occurring for an individual in each category. The odds are defined as the ratio of the probability of the event occurring compared with the probability of it not occurring: if the probability of an event is p, the odds are p/(1-p). The tables in this report show the odds ratio for each category of the independent variables included in the final models. This is derived by dividing the odds for that category by the odds for a defined reference category: this is often the group with the lowest odds so that odds ratios for other groups are greater than 1.0. Where more than one independent variable is included in the model, the odds and odds ratios are adjusted for the effects of all other variables in the model.

Looking again at Table 3.2, we see that the odds ratio for people with a CIS-R score of 18 or more was 3.2. Thus the odds of having consulted a GP were more than three times higher for this group than for people with a low CIS-R score of less than 12. The probability of having consulted a GP will also be higher for this group but it does not follow that there is a three-fold difference in the probabilities for these groups.

The tables indicate which odds ratios are significantly different ($p<0.05$) from 1.0 using conventional markings for the significance level. The results marked in this way are considered to be unlikely to have occurred by chance and so are indicative of a real difference between groups. Because the models reported here have been run on unweighted data, the significance levels should be treated with some caution and used mainly as indicators of the reliability of results. So, looking again at Table 3.2, the greater odds for individuals with a CIS-R score of 18 or more compared with a score of less than 12 is a highly significant result which is very unlikely to have occurred by chance. The difference in odds between people living in private households and those in supported accommodation is, however, a less reliable result.

The models included in this report look only at main effects (ie. the association between being on medication and age) and no interaction terms were fitted (ie. whether the association between age and being on medication varied according to sex). Such interactions were considered to be too detailed for the scope of this report which aims to investigate the main factors associated with a wide variety of dependent measures.

1.6 Explanatory variables used in the analysis

For each of the variables of interest, one logistic regression model was run using all of the independent variables identified as being of interest. These included a variety of socio-demographic variables collected for informants, and indicators of the person's current living arrangements and physical health. The main survey measure of the informant's neurotic symptomatology, the CIS-R score, was also included. The variables were:

- age group - 16-34; 35-44; 45-54; 55-64
- sex - male; female
- marital status - married or cohabiting; single; widowed, divorced or separated
- ethnicity - white; other
- qualifications - GCE A-level or higher; other qualifications; none
- occupation type (based on current or most recent occupation) - non-manual; manual or never worked
- living arrangements - living in supported accommodation; living in private household
- family circumstances - living in supported accommodation; in private household, not with family members; in private household, with members of family
- household type/size - in supported accommodation; one person household; two people in household; three or more people in household

- housing tenure - in supported accommodation; owner occupied; rented

- physical health - reported longstanding physical illness; no physical illness

- CIS-R score (see Appendix A for details). This gives the prevalence of symptoms of neurotic psychopathology derived from 14 sections, each of which covers a particular area of neurotic symptoms. Summed scores from all sections range between 0 and 57. The overall threshold score for significant neurotic morbidity is 12.
  - scores of 0-11; 12-17; 18 or more

It was not possible to include variables based on SCAN data in the analyses as these data were not available for all informants included in this sample of people with a psychotic disorder. SCAN interviews were not carried out for any informants identified through the institutional sample. (See Section 1.4 for full details of the criteria used.)

In a small number of cases, a second stage of analysis was run for selected dependent variables. This used the variables included in the first-stage model together with specific other measures, elsewhere treated as dependent variables. These variables were only included where a causal link with the dependent variable might be hypothesised. Further investigation of the associations between the main dependent measures was carried out separately and is reported in Chapter 8.

1.7 Characteristics of the sample

The main characteristics of the sample are summarised in Tables 1.1 and 1.2. Because of the importance of living arrangements in subsequent analyses in this report, the characteristics of informants living in supported accommodation and those in other, independent, households are shown separately.

There are clearly some differences between the composition of the two parts of the sample. The sample in supported accommodation included a higher percentage of men and more single people than the sample in private households. People in supported accommodation tended to be older and less well qualified and were more likely to be in manual occupation groups. They were also more likely to have a sub-threshold CIS-R score. These differences in distributions for the two main parts of the sample underline the importance of using multi-variate analysis which allows investigation of the association of living arrangements with each dependent variable after having taken account of the effects of other socio-demographic variables.

Notes and references

1 Meltzer H., Gill B., Petticrew M. and Hinds K. (1995) *OPCS Surveys of Psychiatric Morbidity in Great Britain. Report 1: The prevalence of psychiatric morbidity among adults living in private households.* HMSO: London.

2 Meltzer H., Gill B., Petticrew M. and Hinds K. (1996) *OPCS Surveys of Psychiatric Morbidity in Great Britain. Report 4: The prevalence of psychiatric morbidity among adults living in institutions.* HMSO: London.

3 Gill B., Meltzer H., Hinds K. and Petticrew M. (1996) *OPCS Surveys of Psychiatric Morbidity in Great Britain. Report 7: Psychiatric morbidity among homeless people.* HMSO: London.

4 This was coded by interviewers who are familiar with the usual survey definition of a household.

5 Wing, J.K., Babor, T., Brugha, T., Burke, J., Cooper, J.E., Giel, R., Jablensky, A., Regier, D. and Sartorius, N. (1990) SCAN: Schedules for Clinical Assessment in Neuropsychiatry, *Archives of General Psychiatry*, 47, 586-593.
World Health Organisation (1992) *Schedules for Clinical Assessment of Neuropsychiatry*, Division of Mental Health: Geneva.

6 Bebbington, P.E. and Nayani, T. (1995) The Psychosis Screening questionnaire, *International Journal of Methods in Psychiatric Research*, 5, 11-19.

7 Wing, J.K., Nixon, J., Mann, S.A. and Leff, J.P. (1977) Reliability of the PSE used in a population survey, *Psychological Medicine*, 7, 505-516.

Table 1.1 Characteristics of the sample

	Living in private households	Living in supported accommodation	All
	%	%	%
Sex			
Male	47	76	57
Female	53	24	43
Age			
16–34	28	16	23
35–44	23	26	24
45–54	27	32	29
55–64	22	27	24
Marital status			
Married or cohabiting	38	5	26
Single	40	74	52
Widowed, divorced or separated	22	21	22
Ethnicity			
White	96	98	96
Non-white	4	2	4
Qualifications			
A-level or higher	24	15	20
Other	28	24	27
None	48	61	53
Occupation type			
Non-manual	37	22	31
Manual/never worked	63	78	69
Physical illness			
No	62	71	65
Yes	38	29	35
CIS–R score			
0–11	48	68	55
12–17	19	13	17
18 or more	33	19	28
Base	*298*	*172*	*470*

Table 1.2 Characteristics of the sample

	All
	%
Family circumstances	
Supported accommodation	37
Not with family	25
With family	38
Household type/size	
Supported accommodation	37
Informant only	18
2 people	20
3 or more people	26
Housing tenure	
Owned	29
Rented	34
Supported accommodation	37
Base	*470*

2 Medication

2.1 Introduction

The various surveys of psychiatric morbidity collected information about the medication which informants were receiving at the time of interview. The medicines listed were subsequently coded to broad classes based on the British National Formulary.[1] This chapter is concerned with drugs used in the treatment of psychotic disorders and the categories considered are:

- drugs used for psychoses and other conditions including

  antipsychotic drugs
  antipsychotic depot injections
  antimanic drugs

- antidepressants

2.2 Use of medication

Over four fifths of the informants in this sample reported that they were receiving the types of medication considered here. The majority of these were receiving antipsychotic medication. In addition, about a quarter of the sample were on antidepressants. The proportion of informants on antipsychotic or either type of medication varied with the age of the informant and also tended to be higher among those living in supported accommodation. *(Table 2.1)*

Logistic regression analysis was used to investigate which characteristics were most strongly associated with being on medication. Details of this method of analysis are given in Section 1.5. As shown by the results in Table 2.2, four variables were significantly associated with being on either antipsychotic or antidepressant medication - age, ethnic group, whether the informant had a longstanding physical illness and CIS-R score. After allowing for the effect of these variables, living arrangements were not significantly related to the probability of being on medication.

Age was the strongest correlate of being on antipsychotic or antidepressant medication, although the probability did not increase throughout the age range and informants in the 45-54 age group had the highest odds. The relationship between age and being on medication is illustrated in Table 2.1, although the figures shown are not, of course, adjusted for the effects of other significant variables included in the logistic regression model. There is a particularly strong association between age and being on antipsychotic medication: percentages ranged from 70% of the 16-34 age group to 93% of the 45-54 group and 86% of those aged 55-64. CIS-R score was also included as a significant factor in the model although the association was not strong and the pattern was difficult to interpret. Informants with a score of 12-17 on the CIS-R were less likely to be on medication than either those showing fewer or more symptoms of neurosis. The association may be complicated in this case because the type of medication may actually affect the level of neurotic symptoms shown. *(Table 2.2)*

Depot injections

Antipsychotic medication may be given by regular injections at between weekly and monthly intervals. These are often termed depot injections. As the surveys collected detailed information on all medicines received, it was possible to identify which informants were receiving this type of medication and to investigate the characteristics associated with this form of treatment.

In our sample of people with a psychotic disorder, about three in ten of those on antipsychotic medication were receiving depot injections. The prevalence appeared to be higher for the sample of people living in supported accommodation than for those living in private households. However, this did not show through in the multi-variate analysis so there was not an association with living arrangements after having allowed for the stronger associations with the other variables included in the logistic regression model. *(Table 2.3)*

Of the characteristics available for analysis, occupation type and CIS-R score were most strongly associated with being on depot injections rather than oral medication. Informants in manual occupations or who had never worked were more likely than non-manual groups to be receiving depot injections. The reasons for this difference are not clear and it may simply reflect variation in the nature or severity of the psychotic disorders for these groups. In addition to this effect, informants with a CIS-R score of 12-17 were less likely than those below the threshold score of 12 to be receiving injections although the association was relatively weak. This is a similar association to that seen for all medication and, again, there is no strong argument for causality in the analysis. *(Table 2.4)*

2.3 Compliance with medicinal regimes

Informants were asked a number of questions about compliance with the medication that they were receiving and with other treatments which had been suggested for the conditions that they reported. This section looks first at compliance with dosage for those who were on medication for a psychotic condition, and then at whether members of our sample had ever stopped or refused treatment.

Compliance with dosage for antipsychotic medication

One in four adults (27%) in this sample who were on medication for a psychotic condition said that they sometimes did not take the medication prescribed and one in six (16%) said that they sometimes took more than the stated dose. Prevalence was similar for people living in supported accommodation and those in private households. *(Table 2.5)*

About one half of the people who sometimes did not take medication said that on the last occasion they had forgotten to take it. Otherwise, around one in seven mentioned that they did not need it, did not like the side effects or did not like taking drugs. Among those who sometimes took more than the stated dose, most (82%) reported that they had last done this in order to control the symptoms. *(Table 2.6)*

The patterns of association were similar for both types of non-compliance, with the CIS-R score emerging as the sole significant factor in the logistic regression models. In both cases, informants with a CIS-R score of 18 or over were much more likely than those with a lower score to be non-compliant. Because the measure of non-compliance used in this analysis was not confined to recent or frequent incidents, it seems unlikely that non-compliance would affect the CIS-R score measured at the time of interview but rather that causality is in the direction implied by the model. Thus informants who showed more evidence of anxiety and other neurotic symptoms were more likely not to comply with the dosage for their medication, either because of concern about the effects of taking the drugs or simply because of forgetfulness. *(Table 2.7)*

Compliance with other medication or treatment for a psychotic condition

Around one in seven people in the sample reported that they had either refused other medication or treatment for a psychotic condition or had stopped it of their own accord.

Ten per cent of people had refused treatment and a slightly smaller proportion (7%) had decided to discontinue it *(Table 2.8)*. Two reasons were each cited by about half of those who had refused treatment - concern about side effects and not liking the medication. Among those who had discontinued a treatment of their own accord, concern about side effects was clearly the main reason for having stopped.

Having refused or discontinued medication or treatment was strongly associated with the informant's qualification level and, after allowing for this, no other characteristics had a significant effect. People with GCE A-levels or higher qualifications were clearly more likely than informants with no educational qualifications to have refused or stopped treatment. This suggests that people with higher levels of education may be less accepting than others of the advice given to them by health professionals. *(Table 2.9)*

Notes and references

1 *British National Formulary* Number 26, September 1993. A joint publication of the British Medical Association and the Royal Pharmaceutical Society of Great Britain.

Table 2.1 Type of medication by age and by living arrangements

Type of medication	Age				Private household	Supported accommodation	All
	16–34	35–44	45–54	55–64			
	%	%	%	%	%	%	%
Antipsychotic drugs	70	79	93	86	79	90	83
Antidepressants	20	24	25	28	29	17	24
Either antipsychotic or antidepressant drugs	76	85	96	89	84	92	87
Base	*110*	*112*	*137*	*111*	*298*	*172*	*470*

Table 2.2 Characteristics associated with being on medication for a psychotic condition

		Adjusted OR
Age	16–34	1.0
	35–44	1.7
	45–54	6.1 ***
	55–64	2.6 *
Ethnicity	White	4.7 **
	Non-white	1.0
Physical illness	No	2.2 *
	Yes	1.0
CIS–R score	0–11	1.0
	12–17	0.4 **
	18 or more	0.9

* p<0.05 ** p<0.01 *** p<0.001

Table 2.3 Whether on depot injections or other anti-psychotic medication by living arrangements

Base: on antipsychotic medication

Type of medication	Private household	Supported accommodation	All
	%	%	%
Depot injections	25	34	29
Oral medication only	75	66	71
Base	*236*	*154*	*390*

Table 2.4 Characteristics associated with being on depot injections

Base: on antipsychotic medication

		Adjusted OR
Occupation type	Non-manual	1.0
	Manual	2.4**
CIS–R score	0–11	1.0
	12–17	0.4*
	18 or more	0.6

* p<0.05   ** p<0.01   *** p<0.001

Table 2.5 Non-compliance with dosage for medication by age and by living arrangements

Base: on antipsychotic or antidepressant medication

	Age				Private household	Supported accommodation	All
	16–34	35–44	45–54	55–64			
	%	%	%	%	%	%	%
Sometimes do not take medication	37	33	22	19	28	25	27
Sometimes take more than the stated dose	24	20	12	11	18	14	16
Base	*84*	*95*	*131*	*99*	*251*	*158*	*409*

Table 2.6 Reasons for non-compliance with dosage for antipsychotic or antidepressant medication

	Percentage reporting
Reasons for not taking medication	
Forgot	51
Did not need	16
Don't like to take drugs	13
Side effects	12
Other reasons	21
Base (sometimes did not take medication)	*110*
Reasons for taking more than the stated dose	
Needed more to control symptoms	82
Deliberate overdose	6
Other reasons	15
Base (sometimes took more than the stated dose)	*66*

Table 2.7 Characteristics associated with non-compliance with dosage

Base: on medication

		Sometimes did not take	Sometimes took more than dose
		Adjusted OR	Adjusted OR
CIS–R score	0–11	1.0	1.0
	12–17	1.6	1.5
	18 or more	2.3***	3.3***

* p<0.05 ** p<0.01 *** p<0.001

Table 2.8 If had refused or stopped treatment for a psychotic condition by age and by living arrangements

	Age				Private household	Supported accommodation	All
	16–34	35–44	45–54	55–64			
	%	%	%	%	%	%	%
Refused treatment	9	12	7	12	9	11	10
Stopped treatment of own accord	10	4	9	6	7	8	7
Either refused or stopped treatment	16	15	12	14	14	15	14
Base	*110*	*112*	*137*	*111*	*298*	*172*	*470*

Table 2.9 Characteristics associated with having refused or stopped treatment

Base: all cases

		Adjusted OR
Qualifications	A-level or higher	3.4***
	Other	1.8
	None	1.0

* p<0.05 ** p<0.01 *** p<0.001

3 Use of services

3.1 Introduction

This chapter considers the use of health and other services by adults with a psychotic disorder. The analysis covers four categories of services: GP consultations, in-patient episodes, out-patient visits and visits received by people in their homes, for example from a social worker or psychiatrist.

It is probable that the use of some of these services will vary according to the characteristics of informants, such as their age and living arrangements. As in the previous chapter, logistic regression was used to identify which of the characteristics recorded by the survey were associated with service usage. We would also expect that use of services might be related to the severity and time since onset of a person's psychotic condition. This is a difficult concept to define and anyway, because of variation between the surveys in procedures used and in the way in which members of this sample were defined, we did not have access to comparable measures for all sample members. Thus the analysis cannot directly control for the severity of the psychotic disorder and current symptoms. However, some of the variables included in the analysis, notably living arrangements, may themselves be related to severity. The interpretation of results should take account of possible relationships of this type.

The next three sections discuss results for use of the four separate types of service covered in the surveys while Section 3.5 presents a combined analysis and looks at use of any of the main services. The final section deals with services which people had been offered but had turned down.

3.2 GP consultations

This analysis concentrates on consultations for a mental, nervous or emotional problem. Informants in our sample were asked if they had consulted a GP or family doctor either in the past year or in the past two weeks for a physical complaint or for a mental, nervous or emotional problem. Further details were collected about all consultations in the previous two weeks, including whether the informant was satisfied or dissatisfied with it. Consultations recorded include those made in person or by telephone but exclude visits to a hospital.

In the analysis we looked at the likelihood of having consulted for a mental problem both in the previous 12 months and in the two weeks before the interview. Not surprisingly, only a small proportion (12%) of the sample had consulted a GP during the previous two weeks but about three fifths (61%) of informants had consulted about a mental problem in the year before interview. The probability of having consulted a GP was related to age although the direction of the relationship was unusual in that the rate tended to decrease with age, particularly for 55-64 year olds. *(Table 3.1)*

Logistic regression analysis showed that none of the available characteristics were associated with having consulted a GP in the two weeks before interview. As this is only a short reference period, the chances of any individual having consulted their GP will be highly dependent on whether their condition was stable in the few weeks before the interview. This characteristic is not measured by any of the variables in the analysis and consequently no variables have a strong association with the likelihood of having consulted.

Informants who had consulted a GP in the two weeks before interview were asked if they were satisfied with each consultation. Among the small number of people who had consulted for a mental problem in this period, only 1% had felt dissatisfied with a consultation.

Analysis of having consulted a GP in the past year should be less affected by short term changes in condition and so was expected to give a clearer view of factors associated with seeing a GP. The analysis showed that two variables were associated with having consulted a GP, CIS-R score and the informant's living arrangements. Informants living in supported accommodation were less likely than those living in a private household to have consulted. This pattern may occur because people living in supervised accommodation have more ready access to other services and, as discussed later, are more likely to receive domiciliary visits.

The probability of having consulted a GP in the past year was very strongly associated with CIS-R score. Informants scoring above the threshold level of 12 were more likely than those scoring below this level to have consulted a GP, but the odds were particularly high for people with a CIS-R score of 18 or more. *(Table 3.2)*

3.3 In-patient and out-patient episodes

The surveys asked about any in-patient stays during the previous year and about any visits to a hospital or other location for treatment or check-ups as an out-patient. In-patient episodes involved being in hospital for one night or longer for treatment or tests. Out-patient episodes included visits for treatment or tests to hospitals, day hospitals, clinics, private consulting rooms and day centres but excluded visits to the informant's own doctor and periods in hospital. For each episode it was recorded whether the treatment had been for a physical complaint or for a mental or emotional problem and the type of health professional (usually) seen.

Among this sample of people with a psychotic disorder, one fifth had had an in-patient stay and one half had been an out-patient during the previous year for a mental, nervous or emotional problem. Both tended to be less likely among informants living in supported accommodation. As found in earlier reports in this series, people with psychiatric disorders were much more likely than the general population to have had an in-patient stay.[1] The 1993 General Household Survey[2] found that 12% of women and 7% of men aged 16 to 64 years had been an in-patient in the previous year for any reason. *(Table 3.3)*

The likelihood of having had an in-patient episode for a mental problem was significantly associated with both sex and age. In contrast to the pattern for all in-patient stays in the general population, men were slightly more likely than women in our sample to have had an in-patient stay for a mental, nervous or emotional condition. The probability of having been an in-patient was generally higher for younger informants although the relationship was not linear. Informants in the youngest age group (16-34) and those aged 45-54 were more likely than older informants (aged 55-64) to have had an in-patient stay. Having taken account of the effects of age and sex, there was a weak association between living arrangements and having had an in-patient stay, with higher odds for people living in private households. *(Table 3.4)*

The logistic regression model for out-patient visits again shows an association between living arrangements and service use. As with in-patient visits, informants living in supported accommodation were less likely than others to have visited a hospital or clinic as an out-patient but, among the sample living in private households, usage varied with the size of household. People in private households who lived alone were less likely to have made out-patient visits than those living with other people. There was also a weak association with qualification level. *(Table 3.5)*

3.4 Domiciliary visits

Informants were asked whether they had received any visits during the past year from the following health professionals or voluntary workers:

- Community psychiatric nurse
- Occupational therapist
- Social worker
- Psychiatrist
- Home care worker/home help
- Voluntary worker

Overall, two thirds of the adults in our sample (68%) had received domiciliary visits from one or more of these groups although the proportion varied according to living arrangements. More than four fifths (83%) of people in supported accommodation reported at least one type of visit compared with about three fifths (59%) of people living in private households. The type of visits received also differed for the two groups although visits from a psychiatric nurse were most common for both groups (46%). Among people living in private households, about one in seven had been visited by a social worker (16%) and less than one person in ten had received any other type of domiciliary visit. Visits from a social worker were more common among people living in supported accommodation (41%) and more than one in ten people had been visited by a home care worker/home help or by a psychiatrist. *(Table 3.6)*

Because the frequencies of different types of visit varied markedly, our analysis of the factors associated with domiciliary visits did not try to combine all types of visit but instead looked separately at the three most common types. Results for all three models are given in Table 3.7.

The likelihood of receiving a visit from a community psychiatric nurse was not associated with the informant's living arrangements and the only variable found to have a significant effect was whether or not the informant had a physical illness. The probability of having received this type of visit was higher for people who had no physical illness. The association was not, however, particularly strong and there was no obvious explanation for the result.

Living arrangements were significantly associated with having been visited in the past year by a social worker or by a home care worker. In both cases, people living in supported accommodation were clearly more likely than those in private households to have received visits. In the case of visits by a home care worker, the probability also varied by the family circumstances of people living in private households. As would be expected, informants who lived with other members of their family were less likely than those living on their own or with other non-family members to receive visits from a home care worker.

Age was also significantly associated with having been visited by a social worker or home care worker although the direction of the relationships differed. The probability of having been visited by a social worker was highest for informants in the youngest age group (16-34) whereas older informants (aged 55-64) were more likely to have received visits from a home care worker or home help. This variation reflects the difference in the main purpose of visits by these two groups. Apart from the effects of living arrangements and age, the probability of having been visited by a social worker was also higher for informants with a CIS-R score of 18 or more. *(Table 3.7)*

3.5 Use of any services

We have so far looked separately at the factors associated with use of each of the main services and have found some common patterns. However, it may be that use of different types of service is related, either because one type is dependent on another or because one is complementary to another. In this section we therefore look at the overall pattern of service use in the past year for a mental problem.

The proportions of people using each relevant service are shown in Table 3.8. This covers GP consultations, in-patient stays, out-patient visits and domiciliary visits from a psychiatrist or psychiatric nurse. Overall, nine tenths of the sample had used one or more of these services in the past year for a mental problem. Although informants living in supported accommodation were less likely than those in private households to have used some types of service, this was not evident when all services were taken together. *(Table 3.8)*

Three variables were found to be associated with having used any of the services in the past year: ethnicity, having a physical illness, and the measure for neurotic symptoms, the CIS-R score. White informants were more likely than those classified to an ethnic minority group to have used one of these services. This perhaps accords with the result in the previous chapter that White informants were more likely to be on antipsychotic or antidepressant medication. People who did not have a longstanding physical illness in addition to a psychotic disorder also had higher odds of having used some services. This is not easy to explain but may be related to the fact that informants with a physical illness would also be using these services for a physical health problem, which may result in some under-statement of consultations and visits related to their mental or emotional problems. As found in the analysis for GP visits, informants with a high CIS-R score for neurotic symptoms were also more likely to have used any one of the services. *(Table 3.9)*

3.6 People who had refused help or support

Informants were also asked whether they had, in the past year, turned down any help or support from the health professionals described in the section on domiciliary visits. Those who had refused help were asked about what sort of help or service was offered and their reasons for having turned down visits.

About one in ten people in this sample had refused help or support in the past year. All of the different types of support had been turned down by some informants but the most common categories, each turned down by about one fifth of this group, were help from a community psychiatric nurse, a social worker or counselling service, or a psychiatrist. A range of reasons were cited for turning help down of which the most common were that the informant did not want or need help and that he or she did not think that the service offered would actually help them. *(Table 3.10)*

Logistic regression analysis showed that only educational qualifications were associated with having turned help down. Informants with A-levels or higher qualifications were more likely than those with other, lower, qualifications to have refused support or help. This result is generally consistent with the association between qualifications and having refused or stopped treatment, shown in Table 2.9. *(Table 3.11)*

Finally we looked at whether informants had not seen a doctor or other professional about a mental, nervous or emotional problem when either they themselves or other people around them had thought that they should. About one in six of the people in our sample reported that they had not sought help under these circumstances in the past year. Four reasons were each cited by about one in five of this group. They were that the informant:

- didn't think that anyone could help
- didn't think that it was necessary
- thought that it was a problem that they should be able to cope with on their own
- was afraid of the consequences (treatment, tests, hospitalisation, etc)

CIS-R score was the only characteristic significantly associated with having decided not

to see a doctor. Informants with a higher level of neurotic symptoms (CIS-R score of 18 or more) were more likely than other groups to have decided not to see a doctor. This result is not surprising given that CIS-R score is a summary of neurotic symptoms. It is also consistent with the higher probability of non-compliance with dosage for prescribed medication which was reported for this group in Chapter 2.
(Table 3.11)

Notes and references

1 Meltzer H., Gill B., Petticrew M. and Hinds K. (1995) *OPCS Surveys of Psychiatric Morbidity in Great Britain, Report 2, Physical complaints, service use and treatment of adults with psychiatric disorders.* HMSO: London.

2 Foster K., Jackson B., Thomas M., Hunter P. and Bennett N. (1995) *General Household Survey 1993.* HMSO: London.

Table 3.1 GP consultations for a mental, nervous or emotional problem by age and by living arrangements

	Age				Private household	Supported accommodation	All
	16–34	35–44	45–54	55–64			
	%	%	%	%	%	%	%
Consulted a GP in the past 2 weeks	15	13	12	8	13	10	12
Consulted a GP in the past year	65	63	61	54	66	51	61
Base	*110*	*112*	*137*	*111*	*298*	*172*	*470*

Table 3.2 Characteristics associated with having consulted a GP in the past year for a mental condition

		Adjusted OR
Living arrangements	Private household	1.6*
	Supported accommodation	1.0
CIS–R score	0–11	1.0
	12–17	1.7*
	18 or more	3.2***

* p<0.05 ** p<0.01 *** p<0.001

Table 3.3 In-patient stays and out-patient episodes for a mental, nervous or emotional problem by sex and by informant's living arrangements

	Men	Women	Private household	Supported accommodation	All
	%	%	%	%	%
In-patient stay in the past year	23	17	23	15	20
Out-patient episode in the past year	48	54	55	43	50
Base	*270*	*200*	*298*	*172*	*470*

Table 3.4 Characteristics associated with having had an in-patient stay in the past year for a mental, nervous or emotional problem

		Adusted OR
Sex	Male	1.8*
	Female	1.0
Age	16–34	3.2**
	35–44	1.6
	45–54	2.5*
	55–64	1.0
Living arrangements	Private household	2.0*
	Supported accommodation	1.0

* p<0.05 ** p<0.01 *** p<0.001

Table 3.5 Characteristics associated with having made an out-patient visit in the past year for a mental condition

		Adjusted OR
Qualifications	A-level or higher	1.5
	Other	0.7
	None	1.0
Household type/size	Supported accommodation	1.0
	Informant only	0.9
	2 people	2.3**
	3 or more people	1.8*

* p<0.05 ** p<0.01 *** p<0.001

Table 3.6 Type of domiciliary visits in the past year by living arrangements

	Private household	Supported accommodation	All
	%	%	%
Visited in past year by:			
Community psychiatric nurse	47	46	46
Occupational therapist	1	7	3
Social worker	16	41	25
Psychiatrist	4	13	7
Home care worker/ home help	7	29	15
Voluntary worker	4	9	6
Any of the above	59	83	68
Base	*298*	*172*	*470*

Table 3.7 Characteristics associated with having been visited in the past year by a community psychiatric nurse, social worker or home care worker

		Community psychiatric nurse	Social worker	Home care worker
		Adjusted OR	Adjusted OR	Adjusted OR
Age	16–34		2.6**	1.0
	35–44		1.5	1.1
	45–54		1.2	0.8
	55–64		1.0	2.2*
Physical illness	No	1.5*		
	Yes	1.0		
Living arrangements	Private household		1.0	
	Supported accommodation		4.4***	
Family circumstances	Supported accommodation			11.8***
	Not with family			4.5**
	With family			1.0
CIS–R score	0–11		1.0	
	12–17		0.8	
	18 or more		1.7*	

*p<0.05 ** p<0.01 *** p<0.001

Table 3.8 Use of all services in past year by informant's living arrangements

	Private household	Supported accommodation	All
Use of services in past year	%	%	%
Consulted a GP	66	51	61
In-patient stay	23	15	20
Out-patient visit	55	43	50
Domiciliary visit from psychiatrist or community psychiatric nurse	49	49	49
Used any of the above services	90	87	89
Base	*298*	*172*	*470*

Table 3.9 Characteristics associated with having used any services in the past year

		Adjusted OR
Ethnicity	White	4.2*
	Non-white	1.0
Physical illness	No	2.1*
	Yes	1.0
CIS–R score	0–11	1.0
	12–17	1.0
	18 or more	4.7**

* p<0.05 ** p<0.01 *** p<0.001

Table 3.10 Type of help or support turned down in the past year

Base: informants who turned down help

	Percentage reporting
Service or help turned down	%
Community psychiatric nurse	25
Occupational or industrial therapist	18
Social worker or counselling service	20
Psychiatrist	22
Home care worker or home help	8
Voluntary worker	4
Other	20
Base	*51*

Table 3.11 Characteristics associated with having turned down help that was offered or decided not to see a doctor in the past year

Base: informants who (i) turned down help or (ii) decided not to see a doctor

		Turned down help	Decided not to see a doctor
		Adjusted OR	Adjusted OR
Qualifications	A-level or above	1.9	
	Other	0.4	
	None	1.0	
CIS–R score	0–11		1.0
	12–17		1.2
	18 or more		2.8**

* p<0.05 ** p<0.01 *** p<0.001

4 Activities of daily living

4.1 Introduction

Res
any
of c
was
MR
OP
are:
wei

- I
- I
- I
  I
- I
  I
- I
- I
  I
-

If r
wit
wh
prc

The following analysis first investigates which characteristics were associated with having difficulties with each activity of daily living. We then consider the extent to which the same people reported problems with more than one activity and look at the factors associated with having four or more ADL difficulties. It should noted that some areas of functioning would : be expected to apply to individuals living in ported accommodation or group homes and ormants who spontaneously said that any ivity did not apply to them have been cluded from the relevant analyses below (see e numbers in Table 4.1). The function least ely to apply to this sample of people with chosis was practical activities.

ially, Section 4.4 looks briefly at the extent to ich informants who needed help with ivities received it and whether help was ovided by health professionals or voluntary rkers.

2 Difficulties with ADL

o thirds (67%) of this sample of people with chosis reported difficulties with one or more the listed tasks and most of these (61% of the nple) said that they needed help with one or re activities. The proportion of the sample porting each type of ADL difficulty varied dely. People were least likely to have ficulties with personal or medical care)-14%) and most likely to have problems aling with paperwork (46%). One quarter or re of the sample also had difficulties with ch of managing money, household activities d practical activities.

Overall, informants who lived in supported accommodation were more likely than those living in private households to have ADL difficulties, but this was not true of each individual activity. The difference was most marked for medical care and dealing with paperwork.

Most informants who had difficulties with activities of daily living said that they needed help with them. The greatest discrepancies between the percentage of people having difficulties and the percentage needing help were for personal care and using transport: between one half and two thirds of those reporting these difficulties said that they needed help. *(Table 4.1)*

Following the strategy used throughout this report, the factors associated with having difficulties with each of these areas of functioning were investigated using logistic regression analysis. The dependent variables in the analysis were whether informants had difficulties with each individual ADL and the results for the combined samples are shown in Tables 4.2 and 4.3.

Clearly physical illness has a major effect on the ability to carry out many of the tasks covered in the interview. Earlier reports for these surveys have also shown that having a neurotic health problem, as measured by CIS-R score, increases the likelihood of ADL difficulties.[3] These effects also show through in the results presented here.

Physical illness was associated with activities which required physical exertion - household activities, practical activities and using transport - and also with having difficulties with personal care. For each of these, people with a longstanding physical illness were more likely to have difficulties with the activity. The presence of a physical illness was not, however, significantly associated with difficulties with medical care, dealing with paperwork or managing money.

The level of physical exertion involved in different activities may also show through in an association with age, as was seen for practical activities. Informants in the oldest age group (55-64) were more likely than younger people to have difficulties with this ADL, which covers relatively strenuous activities such as gardening and decorating.

As expected, CIS-R score was also a significant factor in most of the models. Informants with a high score for neurotic symptoms (CIS-R score of 18 or more) were more likely than others to have difficulties with all activities other than dealing with paperwork. The associations were particularly strong for using transport and managing money.

The area of activity which showed least relationship with physical illness or neurotic psychopathology was dealing with paperwork. Informants with no qualifications and men were more likely than more qualified informants and women to report difficulties with this activity of daily living. Men were also more likely than women to have difficulties managing money, and qualification level was also associated with difficulties with medical care.

Once the associations with physical illness and neurotic symptoms had been taken into account, the likelihood of having difficulties with an ADL varied according to living arrangements only in the case of medical care and household activities. In both cases, informants living in supported accommodation were more likely to have difficulties with these activities. The apparent difference by living arrangements in the prevalence of difficulties with paperwork (Table 4.1) was therefore not significant after allowing for the effects of sex and qualification level. *(Tables 4.2 and 4.3)*

4.3 Difficulties with a number of activities

Among this sample of adults with psychosis, a substantial number of people had difficulties with more than one activity: just under two fifths (38%) of people had difficulties with two or more and one in six (17%) with four or more activities. The base for these proportions excludes people who said that any of the listed activities did not apply to them.

In order to investigate which groups within our sample had the most difficulties with activities of daily living, we used a threshold of four ADL difficulties for the dependent variable in a logistic regression analysis. Only three characteristics were significantly associated with having four or more ADL difficulties - physical illness, CIS-R score and living arrangements/ housing tenure. As seen for several of the individual activities, informants with a longstanding physical illness and those with a higher level of neurotic symptoms were more likely than other groups to have difficulties with four or more activities. Having taken account of these effects, there was still a weak association with living arrangements. Informants living in supported accommodation and, among those in private households, people in rented accommodation were more likely to have difficulties with several activities.

4.4 Help with ADL

As seen in Table 4.1, most informants who had difficulties with the different ADL went on to say that they needed help. The majority of those who needed help did receive it, in most cases from other members of their family or from friends. Overall, less than one third of people who needed help received it from a health professional or voluntary worker - the groups considered in Section 3.4. It would be expected that the source of help would be related to the informant's living arrangements, particularly whether they were living in supported accommodation and this is confirmed by the pattern of results in Table 4.6, although base numbers for individual ADL are generally small. *(Table 4.6)*

Further analysis was carried out to see whether, having taken account of living arrangements, any other characteristics were associated with receiving help with these activities from health professionals or voluntary workers. In this case the analysis was restricted to the two activities in which the highest proportion of informants had received help from these sources - dealing with paperwork and medical care. The analyses confirmed that in both cases living arrangements were the only significant factor with much higher odds for informants living in supported accommodation than for others. *(Table 4.7)*

Notes and references

1 Brewin C.R. and Wing J.K. (1989) *MRC Needs for Care assessment*, MRC Social Psychiatry Unit, Institute of Psychiatry: London.

2 Martin J., Meltzer H. and Elliot D. (1988) *The OPCS Surveys of Disability in Great Britain, Report 1, The prevalence of disability among adults*, HMSO: London.

3 Meltzer H., Gill B., Petticrew M. and Hinds K. (1995) *OPCS Surveys of Psychiatric Morbidity in Great Britain, Report 3, Economic activity and social functioning of adults with psychiatric disorders*, HMSO: London.

Table 4.1 Difficulties with activities of daily living

Activity	Percentage who had difficulty with activity			Percentage who needed help with activity	*Base (All)*
	Private household	Supported accommodation	All		
Personal care	9	11	10	5	*465*
Medical care	9	22	14	13	*462*
Using transport	23	18	21	14	*464*
Managing money	25	30	27	22	*447*
Household activities	26	34	29	24	*434*
Practical activities	34	27	32	28	*367*
Dealing with paperwork	39	57	46	41	*459*
Any one activity	64	72	67	61	*465*

Table 4.2 Characteristics associated with difficulties with personal care, medical care, using transport and managing money

		Adjusted Odds Ratios for final models			
		Personal care	Medical care	Using transport	Managing money
Age	16–34				
	35–44				
	45–54				
	55–64				
Sex	Female				1.0
	Male				2.2***
Qualifications	A-level or above		1.0		
	Other qualifications		1.8		
	None		3.0*		
Occupation type	Non-manual			1.0	
	Manual/never worked			1.8*	
Living arrangements	Private household		1.0		
	Supported accommodation		3.1***		
Housing tenure	Owned				
	Rented				
	Supported accommodation				
Physical illness	No	1.0		1.0	
	Yes	2.8**		1.8*	
CIS–R score	0–11	1.0	1.0	1.0	1.0
	12–17	1.0	2.1*	2.1*	1.6
	18 or more	2.4*	2.2*	3.3***	2.7***

* p<0.05 ** p<0.01 *** p<0.001

Table 4.3 Characteristics associated with difficulties with household activities, practical activities and dealing with paperwork

		Adjusted Odds Ratios for final models		
		Household activities	Practical activities	Dealing with paperwork
Age	16–34		1.0	
	35–44		1.0	
	45–54		1.8	
	55–64		2.3*	
Sex	Female			1.0
	Male			2.3***
Qualifications	A-level or above			1.0
	Other qualifications			2.0*
	None			6.0***
Occupation type	Non-manual			
	Manual/never worked			
Living arrangements	Private household			
	Supported accommodation			
Housing tenure	Owned	1.0		
	Rented	1.8*		
	Supported accommodation	2.4**		
Physical illness	No	1.0	1.0	
	Yes	2.2***	1.8*	
CIS–R score	0–11	1.0	1.0	
	12–17	1.2	1.7	
	18 or more	2.2**	2.2*	

*p<0.05 ** p<0.01 *** p<0.001

Table 4.4 Number of ADL difficulties

Base: all ADL applicable

	Frequency	Cumulative percentage
Number of ADL difficulties	%	%
4 or more	17	17
3	8	25
2	13	38
1	24	62
None	38	100
Base	*354*	*354*

Table 4.5 Characteristics associated with having difficulties with four or more ADL

		Adjusted OR
Physical illness	No	1.0
	Yes	2.0**
Housing tenure	Owned	1.0
	Rented	1.9*
	Supported accommodation	2.1*
CIS–R score	0–11	1.0
	12–17	1.3
	18 or more	2.6***

* p<0.05 ** p<0.01 *** p<0.001

Table 4.6 Help from a health professional, social or voluntary worker by informant's living arrangements

Base: need help with activity

	Private household	Supported accommodation	All
ADL	*Percentage of those requiring help*		
Personal care	(8)	(50)	(28)
Medical care	(8)	(49)	31
Using transport	10	(38)	17
Managing money	11	33	21
Household activities	12	49	28
Practical activities	4	(32)	11
Dealing with paperwork	16	52	33
Base (Maximum)	*100*	*90*	*190*

( ) Base less than 40

Table 4.7 Characteristics associated with receiving help for ADL difficulties from a health professional, social or voluntary worker

Base: people receiving help for stated ADL difficulty

		Adjusted Odds Ratios	
		Medical care	Dealing with paperwork
Living arrangements	Private household	1.0	
	Supported accommodation	16.7***	
Family circumstances	Supported accommodation		6.5**
	Not with family		2.5
	With family		1.0

*p<0.05 ** p<0.01 *** p<0.001

5 Economic activity and finances

5.1 Introduction

This chapter examines variation in economic activity and income among this sample of people with psychosis. Logistic regression was used to investigate the characteristics associated with three categories of economic activity - working, unemployed and permanently unable to work due to long term sickness or disability. We also look at some of the circumstances of people in these categories, concentrating on the extent to which mental health problems had affected their situation.

Section 5.3 then considers some aspects of the financial circumstances of the sample, particularly receipt of benefits.

5.2 Economic activity

Among the whole population, economic activity is associated with a number of variables, notably sex and age.[1] Overall, half of this sample of people with psychosis were classed as permanently unable to work, about one in five were in employment and one in eight were unemployed. As in the general population, the percentage of people who were economically active (working or unemployed) in our sample decreased with age. Above the age of 55, this was offset by an increase in the proportion of retired people (included as 'other economically inactive'). *(Table 5.1)*

Differences in economic activity by sex were less important for our sample than in the general population. Broadly similar proportions of men and women were working and unemployed although there was a greater difference between men and women in the proportions in each of the economically inactive groups: women are more likely to be classified as keeping house and so included in the 'other economically inactive' category. In addition to differences between men and women, economic activity also varied according to living arrangements. Thus, adults in supported accommodation were less likely than those in private households to be economically active (either working or unemployed) and more likely to be permanently unable to work. *(Tables 5.1 and 5.2)*

In order to explore characteristics associated with economic activity, logistic regression models were run for the probability of working, being unemployed and being permanently unable to work. The results are shown in Table 5.3.

The most consistent feature of the results is the association between economic activity and living arrangements: adults living in supported accommodation were more likely than those in private households to be permanently unable to work, and less likely to be unemployed. The association between living arrangements and being in work is more complicated as the probability differs within the sample in private households. People living in owner occupied accommodation were more likely to be in work than those living either in rented or supported accommodation. In contrast, they were less likely than either of the other two groups to be classed as unable to work. The strong association between housing tenure and economic activity does not, of course, indicate the direction of causality even though economic activity was the dependent variable in our model. It may be, for example, that people who were working and thus earning a regular wage or salary are more likely than others to be able to secure a mortgage.

In addition to the associations with living arrangements and housing tenure, there was also a weaker association between economic activity and ethnicity. The small number of informants classified as non-White had a higher probability of being in work than the majority White group, and were correspondingly less likely to be classed as permanently unable to work.

The probability of being unemployed was significantly associated with qualification level as well as with living arrangements, as already described. Informants with higher educational qualifications were the most likely group to be unemployed. This is the reverse of the relationship seen for the general population[2] and reflects the circumstances of this specific sample.

A larger number of characteristics were associated with being permanently unable to work than with either of the other two categories. The associations with living arrangements, housing tenure and ethnicity were the reverse of those discussed for the model for being in work. In addition, the probability of being unable to work was also related to age, marital status, occupation type and CIS-R score. The probability of being classed as permanently sick was higher for people aged 35 or over than for those aged 16-34 and, having taken this into account, was then higher for single (never-married) people. Informants who had never worked, or were in manual occupations, also had higher odds of being classified as unable to work.

Earlier reports in this series showed that there were clear patterns of differences in economic activity by CIS-R scores. For both men and women, the probability of being in employment decreased with increasing CIS-R score and those with a high score were more likely than those with a low score to be classified as either unemployed or unable to work.[3] Among this sample of people with psychosis, neurotic symptoms were only associated with the likelihood of being permanently unable to work. After allowing for the effects of other variables included in the model, informants with a CIS-R score above the threshold level of 12 were more likely than those scoring below this level to be unable to work. Thus it appears that the existence of significant neurotic symptoms as well as a psychotic condition is associated with a greater probability of being classified as unable to work because of longstanding illness or disability. *(Table 5.3)*

The effect of mental health problems on employment

As well as asking about current economic status, the surveys included specific questions about the extent to which mental health problems had directly affected the informant's employment status and ability to work.

Among the adults in our sample who were in paid employment at the time of the interview, more than half said that their health or the way that they had been feeling had caused them to take time off work in the past year. The average amount of time taken off work in the past year was 42 days. This includes any time taken off due to physical illness.

About one in three of all working adults in our sample worked in sheltered employment, and this rose to around three quarters of those who were living in supported accommodation. The definition of sheltered employment included sheltered places with ordinary employers as well as work schemes run by local authorities or voluntary associations. Among those informants who were not currently working but had previously had a job, about one half said that leaving their job had had something to do with a mental, nervous or emotional problem. Of those who were not currently working but were not retired, two fifths thought that the way they had been feeling made it impossible to do any kind of paid work. A similar proportion thought that they could do work of some kind, including sheltered or part-time work.

5.3 Finances

Survey respondents who were classified as having a mental disorder were asked some basic questions about their financial circumstances. Informants in the sample in institutions were asked if they controlled their own finances and, if not, who exercised this control. Those who did control their own financial affairs and all informants in private households were asked about receipt of state benefits, other sources of income and their gross income.

Basic data on the financial circumstances of the sample are given in Table 5.4. About one in ten of the total sample (one in three people in the institutional sample) did not have control of their own finances. In these circumstances, control was spread among a variety of groups.

Those who had control of their own finances reported receiving a variety of state benefits, of which those most frequently mentioned are shown in the table. Two fifths of the sample were receiving income support or family credit, two fifths were receiving invalidity pension, invalidity benefit or allowance, and about one in seven were receiving disability living allowance. Overall, more than half of the sample were receiving some type of benefit relating to disability.[4] One in six of those asked about their finances were receiving earned income or income from self-employment. *(Table 5.4)*

The median gross weekly income of people in the total sample who had control of their finances was £60-79. This compares with a median weekly gross income of £80-99 found by the psychiatric morbidity surveys[3] for people with a neurotic disorder and £140-£159 for the general population.[2] Not surprisingly, people who had an earned income or salary had a higher median income (£120-139) than others (£60-79).

Logistic regression analysis was used to identify which characteristics were associated with being in receipt of benefits relating to disability.[4] The analysis was run for those members of the sample who had control of their finances, and hence had been asked about receipt of benefits, but excluding those who were in paid employment at the time of the interview.

The variable most strongly associated with receiving these benefits was the informant's sex, although whether he or she had a longstanding physical illness was also a significant factor. In general, men were more likely than women to receive benefits. Clearly the financial position of individuals cannot be entirely separated from that of other members of their household, and marital status would be expected to be related to the odds of receiving benefits. This shows through in the analysis in that the odds were highest for married or cohabiting men and lowest for married or cohabiting women. Having taken account of this effect, the odds of receiving disability-related benefits were higher for people who had a longstanding physical illness in addition to a psychotic condition. The probability of receiving benefits was not independently associated with the informant's living arrangements. *(Table 5.5)*

Notes and references

1 Office of Population Censuses and Surveys, Series LFS no 9, (1992) *Labour Force Survey 1990 and 1991*, HMSO: London.

2 Foster K., Jackson B., Thomas M., Hunter P. and Bennett N. (1995) *General Household Survey 1993*, HMSO: London.

3 Meltzer H., Gill B., Petticrew M. and Hinds K. (1995) *OPCS Surveys of Psychiatric Morbidity in Great Britain. Report 3: Economic activity and social functioning of adults with psychiatric disorders*, HMSO: London.

4 The benefits included in this analysis were: invalidity pension, benefit or allowance; severe disablement allowance; mobility allowance; industrial disablement allowance; attendance allowance; disability living allowance; disability working allowance; invalid care allowance; war disablement pension.

Table 5.1 Economic activity by sex and by age of informant

Economic activity	Men	Women	Age group				All
			16–34	35–44	45–54	55–64	
	%	%	%	%	%	%	%
Working	17	21	25	22	16	13	19
Unemployed	14	12	22	14	11	6	13
Permanently unable to work	58	42	39	52	61	50	51
Other economically inactive	11	26	15	12	12	32	17
Base	*269*	*198*	*109*	*111*	*136*	*111*	*467*

Table 5.2 Economic activity by sex and living arrangements

Economic activity	Private household			Supported accommodation		
	Men	Women	All	Men	Women	All
	%	%	%	%	%	%
Working	19	22	21	15	15	15
Unemployed	20	13	17	8	5	8
Permanently unable to work	50	36	43	66	63	65
Other economically inactive	11	28	20	11	17	12
Base	*139*	*157*	*296*	*130*	*41*	*171*

Table 5.3 Characteristics associated with economic activity status: working, unemployed and unable to work

		Working Adjusted OR	Unemployed Adjusted OR	Unable to work Adjusted OR
Age	16–34			1.0
	35–44			2.5**
	45–54			3.7***
	55–64			2.2*
Marital status	Married or cohabiting			1.0
	Single			2.3**
	Widowed, divorced or separated			1.3
Ethnicity	White	1.0		7.7*
	Non-white	3.4*		1.0
Qualifications	A-level or above		2.7**	
	Other		1.9	
	None		1.0	
Occupation type	Non-manual			1.0
	Manual/never worked			2.2***
Living arrangements	Private household		2.1*	
	Supported accommodation		1.0	
Housing tenure	Owned	4.7***		1.0
	Rented	1.0		3.3***
	Supported accommodation	1.6		3.9***
CIS-R score	0–11			1.0
	12–17			2.2**
	18 or more			3.0**

* p<0.05 **p<0.01 ***p<0.001

Table 5.4 Financial circumstances of members of the sample

	All
	%
Financial control	
Controls own financial affairs	89
Institutional sample and someone else has financial control	11
Base	*470*
Percentage receiving:	
Income support or family credit	38
Invalidity pension, benefit or allowance	42
Disability living allowance	15
Any disability-related benefit	56
Earned income/income from self-employment	17
Base (controls own finances)	*417*

Table 5.5 Characteristics associated with receiving benefits related to disability

Base: has control of own finances, not in employment

		Adjusted OR
Sex and marital status	Married/cohabiting woman	1.0
	Unmarried woman	1.7
	Married/cohabiting man	5.6**
	Unmarried man	3.8***
Physical illness	No	1.0
	Yes	1.7*

*p<0.05 **p<0.01 ***p<0.001

6 Social functioning

6.1 Introduction

The psychiatric morbidity surveys considered three aspects of social functioning:

- extent of social networks
- self-perceived social support
- involvement in social and leisure activities

This chapter looks at each aspect of social functioning in turn although Section 6.3 also considers the relationship between primary support group and perceived social support. Further analysis of the inter-relationship between these and other measures can be found in Chapter 8.

6.2 Extent of social networks

Information about social networks focused on the numbers of friends and relatives (aged 16 and over) respondents felt close to. Because of the personal nature of the questions, they were not asked of proxy respondents. Data were collected about three groups of people:

1) adults who lived with respondents who they felt close to;

2) relatives living elsewhere who respondents felt close to;

3) friends or acquaintances living elsewhere who respondents would describe as close or good friends.

Close friends and relatives, including people from all of the groups listed above, form an individual's 'primary support group'. Previous research has suggested that adults with a total primary support group of three people or fewer are at greatest risk of psychiatric morbidity[1] and this relationship was confirmed by earlier results from these surveys.[2]

For the analysis in this report, the size of the primary support group was divided into three bands, 0-3, 4-8, and 9 or more, and the dependent variable in the logistic regression was whether the respondent had a network of 0-3 people or not. About 2% of the sample who were interviewed in person gave no answer to one or more of the constituent questions needed to calculate the size of primary support group and hence they have been excluded from the analyses presented here.

Among our sample of people with psychosis, about one quarter (26%) felt close to fewer than four people: this compares with a figure of 7% for all people sampled in the private household survey.[2] Informants living in supported accommodation were more likely than those living in private households to have only a small network of close family and friends: almost a third (31%) had a primary support group of three or fewer people. *(Table 6.1)*

Multiple logistic regression was carried out to test whether there was a significant association between living arrangements and having a small primary support group after having taken account of the effects of other variables. The analysis identified a difference among those living in private households as well as the higher odds for those living in supported accommodation, but no other characteristics were significantly associated with having a small social network. Not surprisingly, informants who lived with other members of their family were less likely to have a primary support group of three or fewer people than were informants who either lived on their own or with other people who were not family members.

As a second stage of analysis we tested whether, having allowed for the effects of living arrangements, the size of social network was affected by the informant's economic activity status. The analysis clearly showed that sample members who were classed as unable to work because of longstanding illness or disability were more likely than those who were working to have a small social network: this was in addition to the association with living arrangements already mentioned. There was, however, no significant difference between those who were in work and those who were either unemployed or in other economically inactive categories. *(Table 6.2)*

6.3 Perceived social support

Perceived social support was assessed from respondents' answers to seven questions taken from the 1987 Health and Lifestyle Survey.[3] The questions take the form of statements which individuals could say were not true, partly true, or certainly true of their family and friends, giving scores of 1, 2 or 3 respectively for each question. The questions were:

There are people I know - amongst my family or friends

- who do things to make me happy
- who make me feel loved
- who can be relied on no matter what happens
- who would see that I am taken care of if I needed to be
- who accept me just as I am
- who make me feel an important part of their lives
- who give me support and encouragement

Overall scores ranged from 7 to 21 and these were categorised into three groups, following the conventions used in earlier reports in this series:

- score of 21 - no lack of social support
- 18 to 20 - moderate lack of social support
- 17 or less - severe lack of social support.

In this sample of people with psychosis, one third were classified as having a severe lack of social support: this compares with an estimate of around 14% for the total population living in private households. Informants living in supported accommodation were more likely than those in private households to fall into this category: the proportion was about one half for the former group and one quarter for the latter. *(Table 6.3)*

Logistic regression analysis confirmed the strong association between living arrangements and perceiving a severe lack of social support. As was also the case for size of primary support group, an important factor for informants in private households was whether the informant was living with members of his or her family or not: those living on their own or with other unrelated people were more likely to feel that they lacked support. In addition to the effects of living arrangements, people with a greater level of neurotic symptoms were more likely than others to have a severe lack of support. Given that the social support measure is about perception rather than being an objective measure, it is perhaps not surprising that people who showed evidence of being anxious or depressed might also be more likely to think that they lacked support from other people.

As a second stage of analysis, the size of primary support group was entered into the model in order to test whether perceived social support was related to the size of an informant's social network. The two measures of social support differ because the latter is an objective measure of size whereas the former is a more subjective measure of the quality of support, so the relationship between them would not necessarily be strong. Nonetheless, it might be that people with only a small number of close relatives or friends would also feel less certain that they could rely on anyone to give support of the types covered by the questions.

The results showed that the odds of perceiving a severe lack of social support were significantly higher for people with a small social network. The earlier associations between living arrangements and CIS-R score were largely independent of this association with size of support group. Thus, after taking account of the relationship between size of primary support group and perceived social support, informants living in supported accommodation were still more likely than others to perceive a severe lack of social support. This may be because people living in institutional households have a primary network mainly comprising their carers and other people in a similar situation to their own. These relationships may be perceived as offering less support than do those with relatives and chosen friends. *(Table 6.4)*

6.4 Social and leisure activities

Survey respondents were asked about the activities that they did, both in and out of the home, during their leisure time. The lists of activities used in the survey were developed specifically for this project: nine activities in the home and fifteen outside the home were included.

Activities in and around the home
Entertaining friends or relatives
Writing letters/telephoning
Reading books and newspapers
Watching TV/listening to the radio
Listening to music
Hobbies (including art and crafts, knitting, playing a musical instrument, writing poetry)
Gardening
DIY/car maintenance
Games (including cards, computer games, betting and gambling)

Activities out of the home
Visiting friends or relatives
Pubs, restaurants
Night clubs, discos
Clubs, organisations
Classes, lectures
Going for a walk/walking the dog
Sports (including keep fit, cycling, swimming, football and horse riding)
Sports as a spectator
Cinema, theatre, concerts
Bingo, amusement arcades
Bookmakers, betting and gambling
Shopping
Church
Political activities
Library

This section looks at the total number of activities that respondents participated in. In earlier analysis of leisure activities in Report 3, a key threshold of four activities was defined and this has been adopted here.

Most of the people in our sample were involved in between four and eight of the 24 listed activities.

About one in eight of the sample were involved in three or fewer activities and the proportions were similar for adults living in private households and those in supported accommodation. *(Table 6.5)*

The dependent variable used in the logistic regression analysis was whether respondents took part in fewer than four activities or not. This analysis confirmed that participating in only a limited range of activities was not associated with the informant's living arrangements but two other variables, educational level and sex, were associated. Of these, the association with qualification level was stronger. Informants with no educational qualifications were less likely than those with A levels or higher qualifications to take part in a broad range of leisure activities. In addition, men participated less than women. *(Table 6.6)*

Attendance at day centres, clubs and training centres

A small number of questions on attendance at specific types of club or centre were also asked in the surveys. The first category considered here is people who attended any one of a range of centres including day centres, clubs for people with physical or mental health problems and social clubs for social activities. The second category is whether informants regularly went to an Adult Education or Training Centre.

Two fifths of people in the sample attended a club or day centre for social activities and the likelihood of doing so was greater for informants living in supported accommodation than those in private households. A much smaller proportion of the sample (7%) regularly went to an Adult Education or Training Centre. In contrast to attendance at a day centre or social club, living arrangements were not associated with attendance at these Centres. The only variable of those tested which had a significant effect was qualifications: informants with the highest qualifications, A levels or above, were more likely than others to be attending centres for further training. *(Tables 6.5 and 6.7)*

Notes and references

1 Brugha T.S., Wing J.K., Brewin C.R., MacCarthy B. and Lesage A. (1993) The relationship of social network deficits with deficits in social functioning in long-term psychiatric disorders, *Social Psychiatry and Psychiatric Epidemiology*, 28, 218-224.

2 Meltzer H., Gill B., Petticrew M.and Hinds K. (1995) *OPCS Surveys of Psychiatric Morbidity in Great Britain. Report 3: Economic activity and social functioning of adults with psychiatric disorders*, HMSO: London.

3 *The Health and Lifestyle Survey*, Health Promotion Research Trust, 1987.

Table 6.1 Size of primary support group by informant's living arrangements

Size of primary support group	Private household	Supported accommodation	All
	%	%	%
0–3 people	24	31	26
4–8 people	44	39	43
9 or more people	32	31	31
Base	*291*	*147*	*438*

Table 6.2 Characteristics associated with having a primary support group of 3 people or fewer

		Stage 1 model Adjusted OR	Stage 2 model Adjusted OR
Living arrangements	Supported accommodation	2.0**	1.8*
	Not with family	2.3**	2.2**
	Living with family	1.0	1.0
Economic activity	Working		1.0
	Unemployed		1.7
	Permanently unable to work		2.6 **
	Other economically inactive		1.6

*p<0.05 **p<0.01 ***p<0.001

Table 6.3 Perceived social support by informant's living arrangements

Perceived social support	Private household	Supported accommodation	All
	%	%	%
Severe lack	25	49	33
Moderate lack	33	19	29
No lack	42	32	39
Base	*281*	*139*	*420*

Table 6.4 Characteristics associated with having a severe lack of perceived social support

		Stage 1 model Adjusted OR	Stage 2 model Adjusted OR
Family circumstances	Supported accommodation	4.7***	4.2***
	Not with family	2.2**	1.7
	Living with family	1.0	1.0
CIS–R score	0–11	1.0	1.0
	12–17	1.2	1.0
	18 or more	2.2**	2.0*
Primary support group	0–3		7.6***
	4–8		2.0*
	9 or more		1.0

*p<0.05 **p<0.01 ***p<0.001

Table 6.5 Number of leisure activities by informant's living arrangements

Number of leisure activities	Private household	Supported accommodation	All
	%	%	%
0–3	11	15	13
4–8	59	66	61
9 or more	30	19	26
Attends club or day centre	31	56	40
Attends Adult Education or Training Centre	7	6	7
Base	*298*	*172*	*470*

Table 6.6 Characteristics associated with taking part in fewer than four activities

		Adjusted OR
Sex	Male	1.8*
	Female	1.0
Qualifications	A-level or above	1.0
	Other	2.3
	None	3.7**

*p<0.05 **p<0.01 ***p<0.001

Table 6.7 Characteristics associated with regularly attending:
a) a day centre, club for people with health problems, or a social club
b) an Adult Education or Adult Training Centre

		Social club	Adult Education Centre
		Adjusted OR	Adjusted OR
Qualifications	A-level or above		3.4**
	Other		1.5
	None		1.0
Family circumstances	Supported accommodation	3.5***	
	Not with family	1.6	
	Living with family	1.0	

*p<0.05 **p<0.01 ***p<0.001

7 Tobacco, alcohol and drugs

7.1 Introduction

Measures of cigarette smoking, alcohol use and drug misuse are available for respondents who had personal interviews.

Drug misuse includes the use of illegal drugs such as cannabis, stimulants and hallucinogens, and the extra-medical use of prescription medicines. The consumption of prescribed medication in general was considered in Chapter 2 of this report.

Obtaining information about people's drinking, drug-taking and cigarette smoking is difficult. Social surveys of the general population consistently record lower levels of alcohol consumption and cigarette smoking than would be expected from alcohol and tobacco sales. This is for a variety of reasons, such as underestimation of amounts drunk at home, poor recall and non-response bias. Another factor, particularly regarding drug misuse, may be respondents' concerns about confidentiality. More discussion of these problems is found in Report 3 of the private household survey.[1] It is not known whether, or to what extent, similar problems occur with regard to data collected in institutions.

7.2 Cigarette smoking

All informants, except proxies, were asked if they had ever smoked cigarettes and if they smoked nowadays. This did not include cigar and pipe smoking. If they did smoke, they were asked how many cigarettes they usually smoked a day on weekdays and at weekends. The questions asked were identical to those in the 1992 General Household Survey (GHS)[2] and the 1993 Health Survey for England,[3] although neither of these surveys covered people living in institutions. For the analysis, adults were grouped into categories depending on whether they had ever smoked, and the average daily amount smoked.

Cigarette smoking categories

Never regularly smoked

Ex-smokers

Current smokers:

Light	Less than 10 a day
Moderate	Less than 20 but more than 10 a day
Heavy	20 or more a day

Among this sample of people with psychosis, around three fifths (58%) were regular smokers and almost two fifths (38%) were classed as heavy smokers (smoking 20 or more cigarettes a day). Both of these proportions are substantially higher than those found by this survey for the general population living in private households: Report 3 showed that 32% of adults were regular smokers and 11% were heavy smokers.[1] In this sample, smoking appeared to be more common among people living in supported accommodation, although prevalence was still above 50% for people in private households. Only one in six (16%) of the sample living in supported accommodation had never regularly smoked; the proportion was roughly double this among people in private households. *(Table 7.1)*

It is well known that cigarette smoking varies by age and sex. In general, men are more likely to have ever smoked regularly and the prevalence of smoking is higher among younger age groups.[2,3] Variation in smoking prevalence by sex was also apparent for this sample (see Table 7.1). The possible association between smoking and age in this sample was investigated by means of logistic regression analysis so that the effects of other possible correlates could also be taken into account.

Two logistic regression analyses were carried out, the first distinguishing between regular smokers and non-smokers, and the second distinguishing between heavy smokers and all others. In the first model, two factors were found to be significantly associated with the probability of being a regular smoker - living arrangements/housing tenure and occupation type. Informants in supported accommodation were most likely to be regular smokers but, among the sample in private households, the odds of being a regular smoker were greater for those living in rented as opposed to owner occupied accommodation. The association between smoking and occupation type was similar to that seen in the general population, with people in manual social class groups more likely to be smokers. Once these factors had been taken into account, there was no significant association between smoking and either age or sex.

The relationship between heavy smoking and living arrangements/housing tenure was similar to the association already described for any level of smoking. Members of the sample living in supported accommodation were again most likely to be heavy smokers. Occupation type was not, however, included in this model but was replaced by age and whether the informant had a longstanding physical illness. Informants in the oldest age group covered (55-64) were significantly less likely than younger groups to be heavy smokers, as also were informants who did not have a longstanding physical complaint. The latter relationship may be due in part to the possibility that some people give up smoking or reduce the number of cigarettes smoked because of ill health. *(Table 7.2)*

7.3 Alcohol consumption

The methodology used on this survey to categorise alcohol consumption is the same as that used by the General Household Survey (GHS)[2] and the 1993 Health Survey for England.[3] Informants were asked how often they had drunk each of the following five types of drink in the previous year and how much of each type they usually drank on any one day:

- Shandy (excluding bottles or cans which have very low alcohol content)
- Beer, lager, stout, cider
- Spirits or liqueurs
- Sherry or martini
- Wine

Informants described their consumption in terms of standard measures which contained similar amounts of alcohol, one unit of alcohol being approximately equivalent to a half pint of beer, a single measure of spirits (1/6 gill), a glass of wine (about 4.5 fluid ounces) or a small glass of sherry or fortified wine (2 fluid ounces).

The alcohol consumption rating is calculated by multiplying the number of units of each type of drink consumed on a 'usual' day by a conversion factor relating to the frequency with which it was drunk, and totalling across all drinks:

Multiplying factors for converting drinking frequency and number of units consumed on a usual day into a number of units consumed per week.

Drinking frequency	Multiplying factor
Almost every day	7.0
5 or 6 times per week	5.5
3 or 4 times per week	3.5
Once or twice per week	1.5
Once or twice per month	0.375 (1.5/4)
Once or twice per 6 months	0.058 (1.5/26)
Once or twice per year	0.029 (1.5/52)

When the survey took place in 1994 the recommended sensible drinking levels were 21 units per week for men and 14 for women. Consumption above these levels was thought to be associated with increased health risks.[4] Although these levels were increased at the end of 1995, analysis of the data in this report is based on the 1994 guidelines to retain comparability with the data from the private household survey and the General Household Survey. Thus, residents who are described as being 'Fairly heavy' 'Heavy' or 'Very heavy' drinkers were consuming over the recommended sensible level of alcohol when the survey took place. The way in which the descriptive labels used here relate to units of alcohol is shown below:

Alcohol consumption categories, based on usual weekly consumption (units) over the previous 12 months

Abstainer	Informant drank no alcohol in the past year	
Occasional drinker	Under 1 unit per week	
	Men (units per week)	Women
Light	1 - 10	1 - 7
Moderate	11 - 21	8 - 14
Over the recommended sensible level:		
Fairly heavy	22 - 35	15 -25
Heavy	36 - 50	26 - 35
Very heavy	51 or more	36 or more

Table 7.3 shows that about two fifths of this sample of people with psychosis either abstained from drinking alcohol or drank only occasionally. This compares with about one fifth (20%) of all people living in private households.[1] Most of those who drank more than one unit on average per week were classed as light drinkers and only about one in ten of the sample drank more than the recommended sensible level per week.

Among the general population in private households, men are significantly more likely than women to drink more than the recommended sensible level and they are less likely to be abstainers or occasional drinkers. This difference in drinking behaviour by sex was also seen in this sample: one third of men were classed as abstainers or occasional drinkers compared with one half of women. There was no clear difference in alcohol consumption by living arrangements. *(Table 7.3)*

Two logistic regression models were used to investigate the characteristics associated with alcohol consumption. The dependent variable in the first model was whether the informant was a regular drinker as opposed to being an abstainer or occasional drinker, and in the second whether the informant drank more than the recommended sensible level or not.

The logistic regression models emphasised the strong associations between drinking and sex and age. Men and younger informants were more likely than women and people aged 45 or over to be regular drinkers. There was not a significant association between age and drinking more heavily and, in this case, the sex of the informant was the only significant effect. As already seen in Table 7.3, men were more likely than women to drink more than the recommended sensible amount. Although Report 3 of this series showed that abstaining and occasional drinking increased with CIS-R score among the general household population, neurotic symptoms were not associated with alcohol consumption level among this sample of people with psychosis. As expected from the comparison of distributions in Table 7.3, alcohol consumption was not associated with the informant's living arrangements. *(Table 7.4)*

7.4 Drug use

Informants were asked about their use of drugs including sedatives, tranquillisers, cannabis, amphetamines, cocaine, heroin, opiates, hallucinogens, ecstasy and glue. In the case of prescribed drugs, most usually sedatives and tranquillisers, we were interested only in their extra-medical use.

Extra-medical and illicit use was ascertained by presenting informants with a list of drugs and asking if they had used any of these without a prescription or more than was prescribed for them or to get high. Sedatives and tranquillisers were placed at the top of the list to deter people who did not use illicit drugs but did misuse medication from assuming the questions did not apply to them.

The questions on drug use and the drug categories were drawn from the drugs section of the Diagnostic Interview Schedule (DIS)[5] and were used in the US ECA study.[6] Questions on injecting drugs and needle sharing were added.

Informants who reported using any of the drugs listed either without a prescription or at more than the prescribed dosage or to get high were then asked if they had taken the drug more than five times in their life. Those who had done so and had also taken the drug in the past twelve months were defined as users of the drug.

The types of drugs used are presented in the following broad groupings owing to the small numbers of people who had taken most types.

Drug categories used in analysis

Sleeping tablets Tranquillisers	Hypnotics
Cannabis	Cannabis
Amphetamines Cocaine/crack	Stimulants
Hallucinogens/ psychedelics Ecstasy	Hallucinogens including Ecstasy
Heroin Opiates Solvents, inhalents	Other drugs

One in ten people in this sample reported extra-medical or illicit use of at least one of the listed drugs in the past year. This compares with about one in twenty people in the general household population. Cannabis (6%) and hypnotics (5%) were the most commonly used drugs. Prevalence of use of all drugs was similar for men and women and did not vary according to living arrangements. *(Table 7.5)*

Because of the small proportions of people taking most of the categories of drugs, logistic regression analysis was restricted to the two most commonly used types - hypnotics and cannabis. A further model was run to identify characteristics associated with use of any of the listed drugs. The patterns of association for use of hypnotics and use of cannabis differed markedly. Use of hypnotics, which covers extra-medical use of sleeping tablets and tranquillisers, was significantly more likely among informants with a CIS-R score of 18 or more and, to a lesser extent, among those with no longstanding physical illness. It is possible that the association with CIS-R score may also reflect an association with the prescribed use of these drugs. Cannabis use was only associated with age. People in the 16-34 age group were more likely than those aged 35 or over to have used the drug in the previous year. In fact, no-one aged 45 or over in this sample reported having used cannabis. The analysis for characteristics associated with use of any drug reflects the strong age-effect for use of cannabis and age is the only significant factor in the model. None of the models identified living arrangements as being significantly associated with drug use and there was a less strong association between drug use and CIS-R score than that described in Report 3 for the general household population. *(Table 7.6)*

Notes and references

1 Meltzer H., Gill B., Petticrew M. and Hinds K. (1995) *OPCS Surveys of Psychiatric Morbidity in Great Britain. Report 3: Economic activity and social functioning of adults with psychiatric disorders*, HMSO: London.

2 Thomas M., Goddard E., Hickman M. and Hunter P. (1994) *1992 General Household Survey*, HMSO: London

3 Bennett N., Dodd T., Flatley J., Freeth S. and Bolling K. (1995) *Health Survey for England, 1993*. HMSO: London.

4 *Alcohol and the heart in perspective: sensible limits reaffirmed*. Report of a joint working group of the Royal College of Physicians, the Royal College of Psychiatrists and the Royal College of General Practitioners. June 1995.

5 Robins L.N., Helzer J.E., Croughan J. and Ratcliff J.S. (1981) National Institute of Mental Health Diagnostic Interview Schedule: Its History, Characteristics and Validity, *Archives of General Psychiatry*, vol 38, pp 381-389.

6 *Psychiatric Disorders in America, The Epidemiologic Catchment Area Study*, edited by Robins LN and Reiger DA (1991) Free Press: New York.

Table 7.1 Cigarette smoking by sex and by living arrangements

Cigarette smoking	Men		Women		Private household		Supported accommodation		All	
	%		%		%		%		%	
Never regular	20		33		30		16		26	
Ex-regular	17		16		18		13		16	
Light	5		4		6		2		5	
Moderate	14	64	16	51	14	52	17	71	15	58
Heavy	44		31		31		52		38	
Base	*248*		*190*		*289*		*149*		*438*	

Table 7.2 Characteristics associated with being a regular smoker and with being a heavy smoker

		Regular smoker Adjusted OR	Heavy smoker Adjusted OR
Age	16–34		2.1*
	35–44		2.0*
	45–54		2.7**
	55–64		1.0
Physical illness	No		1.9**
	Yes		1.0
Housing tenure	Owned	1.0	1.0
	Rented	2.1**	2.4**
	Supported accommodation	3.1***	3.8 ***
Occupation type	Non-manual	1.0	
	Manual	1.7**	

*p<0.05 **p<0.01 ***p<0.001

Table 7.3 Alcohol consumption level by sex and by living arrangements

Alcohol consumption level	Men		Women		Private household		Supported accommodation		All	
	%		%		%		%		%	
Abstainer	18	34	23	50	18	40	25	43	20	41
Occasional drinker	16		27		22		18		21	
Light	36		32		32		39		34	
Moderate	16		11		16		10		14	
Fairly heavy	6		5		6		4		5	
Heavy	3	13	1	7	2	12	1	8	2	11
Very heavy	4		2		3		3		3	
Base	*248*		*190*		*290*		*148*		*438*	

Table 7.4 Characteristics associated with being a regular drinker and with drinking more than the recommended sensible level

		Regular drinker Adjusted OR	Heavier drinker Adjusted OR
Sex	Male	1.9**	1.9*
	Female	1.0	1.0
Age	16–34	4.0**	
	35–44	3.1***	
	45–54	1.9*	
	55-64	1.0	

*p<0.05 **p<0.01 ***p<0.001

Table 7.5 Use of drugs by sex and by living arrangements

Drug type	Men	Women	Private household	Supported accommodation	All
	%	%	%	%	%
Cannabis	6	5	6	5	6
Stimulants	1	1	1	1	1
Hallucinogens inc. Ecstasy	1	1	0	1	1
Hypnotics	5	6	5	6	5
Other drugs	1	–	1	–	0
Any drug	11	10	11	10	11
Base	*270*	*200*	*298*	*172*	*470*

Table 7.6 Characteristics associated with drug use

		Hypnotics Adjusted OR	Cannabis Adjusted OR	Any drug Adjusted OR
Age	16–34		8.6***	17.7***
	35–44			7.1*
	45–54		1.0 (35–64)	3.4
	55–64			1.0
Physical illness	No	2.8*		
	Yes	1.0		
CIS–R score	0–11	1.0		
	12–17	2.3		
	18 or more	6.1***		

*p<0.05 **p<0.01 ***p<0.001

8 The circumstances of people with psychosis

8.1 Introduction

So far this report has looked in turn at use of medication, use of services and at some of the circumstances of people with psychosis, including difficulties with activities of daily living, economic activity, social functioning, and use of tobacco, alcohol and drugs. Having investigated the sample characteristics which are associated with each of these, some strong themes have emerged. These include the differing circumstances and service use of people living in supported accommodation as opposed to those living in private households, and the associations between CIS-R score and several of the variables considered.

In general, the analysis presented so far has not sought to identify associations between the different circumstances used as dependent variables in the models because the aim was to show how the probability of being in a particular category varied with the informant's characteristics. The variables describing informants' circumstances were used as independent variables in other models only in a small number of cases where it was thought that there might be a causal relationship between them.

The aim of this chapter is to move away from the separate consideration of informants' circumstances and to consider the inter-relationships between the different topics covered in earlier chapters. Section 8.2 adopts the pattern followed elsewhere in this report by using logistic regression to investigate inter-relationships. Section 8.3, in contrast, takes a qualitative approach and presents a small number of case studies of informants. These do not aim to be representative of the sample as a whole but are useful in giving a more detailed picture of the circumstances of particular informants in order to illustrate some of the inter-relationships identified by multi-variate analysis on data from the whole sample.

8.2 Identifying groups with different circumstances

The aim of this analysis was to identify relationships between the main topics covered in earlier chapters. This was done by using logistic regression to test for an association between each pair of measures. The dependent variables, listed at the head of the columns in Table 8.1, were all dichotomous and were as used in earlier chapters of this report:

- informant was on antipsychotic or antidepressant medication
- used any of the main health services in the past year (see Section 3.5)
- had difficulties with 4 or more activities of daily living
- classified as unable to work because of longstanding illness
- primary support group of 0-3 people
- severe lack of perceived social support
- took part in 0-3 leisure activities
- heavy smoker (smoked 20 or more cigarettes per day)
- alcohol consumption above the recommended sensible level
- had used illegal drugs

The independent variables covered the same measures but more categories were allowed. These are listed on the left hand side of Table 8.1 and define the rows in the table. Two further independent variables which were prominent in earlier analyses are also included at the foot of the table: the informant's living arrangements and the CIS-R score.

Unlike previous models discussed in this report, the results reflect the association of a single independent variable (row) with the relevant dependent variable (column). Odds ratios are not given but the results are summarised by showing whether the odds for the category were significantly different from those for the reference category. Most of the significant odds ratios were greater than 1.0, indicating a positive relationship between the variables: those which were significantly less than 1.0 are shown in brackets.

Looking across the columns it is apparent that many of the measures used as dependent variables in earlier analyses were strongly associated with each other. The variables which showed least relationship with the other measures considered were alcohol consumption and extra-medical or illegal use of drugs. Drinking more than the recommended sensible level and use of illegal drugs were associated only with heavy cigarette smoking.

A number of strong inter-relationships can be identified from the table. The strongest sets of associations were as follows:

- being on antipsychotic or antidepressant medication and having used some health services in the past year;

- having four or more ADL difficulties and being classified as permanently unable to work;

- having a small primary support group, a severe lack of perceived social support and being involved in few leisure activities.

Further associations can be identified between some of the individual measures describing peoples' social and economic circumstances (as covered in Chapters 4 to 6). So, for example, having several ADL difficulties was associated with participation in few leisure activities, and being classified as unable to work was also associated with the indicators of low social support and having taken part in few leisure activities. These variables which describe informants' circumstances were less clearly associated with medication and use of health services although there is some evidence that people classified as permanently unable to work were more likely than other groups to be on medication and to have used some health services in the previous year.

Generally, the health-related behaviours covered in Chapter 7 were not strongly associated either with measures of informants' social and economic circumstances or their contact with health services. This was particularly the case for alcohol consumption which was only associated with cigarette smoking; people who drank alcohol regularly were more likely than abstainers and occasional drinkers to be heavy smokers. Heavy smoking was more closely related to other variables. In addition to the associations with drinking and drug use, heavy smoking was also associated with being on prescribed medication and being classified as unable to work.

Some interesting associations with illegal drug use emerge when use of cannabis is separated from use of other drugs. Looking across the relevant row of Table 8.1 shows that people who had used cannabis were more likely than others to be heavy smokers or to have several ADL difficulties, whereas they were less likely than other groups to be on prescribed medication. For this sample there was no association between use of illegal drugs and use of health services.

As well as looking at inter-relationships between the various key measures used in earlier analysis, Table 8.1 also summarises the simple associations between these measures and

CIS-R score and living arrangements: these associations do not take account of other possible associations. Five of the ten measures were associated with CIS-R score: informants with a CIS-R score of 18 or more were more likely to have used health services, have 4 or more ADL difficulties and to have used illegal or extra-medical drugs. The relationship between CIS-R score and being on medication was discussed more fully in Chapter 2. Neurotic symptomatology was not significantly associated with the measures of social functioning or with smoking and drinking behaviour.

Living arrangements were strongly associated with three of the measures. Thus, informants in supported accommodation were more likely than those in private households to be classified as unable to work, to have a severe lack of social support, or to be heavy smokers. There was also a weaker association in the same direction between living arrangements and being on medication. *(Table 8.1)*

A second stage of analysis was carried out in order to investigate the characteristics associated with three specific combinations of the dependent measures.

i. Difficulties with 4 or more ADL and classified as unable to work

As already seen, there was a strong association between having difficulties with activities of daily living and being classified as unable to work because of long-term sickness. The combination of these characteristics might be thought of as defining the group with the most severe health problems and impairment, either because of their mental condition alone or because of a combination of mental and physical symptoms.

Multi-variate analysis showed that the probability of having this combination of circumstances was related to living arrangements/housing tenure and to CIS-R score. The odds of showing more severe impairment were higher for informants living in supported accommodation and, among those in private households, for people living in rented accommodation. As mentioned in Section 5.2, the higher probability for those in rented accommodation presumably reflects the greater difficulties in taking on a mortgage for people not in permanent employment. The odds also increased with CIS-R score, with a marked difference between groups with scores below and above the defined threshold level of 12. As in other analyses throughout this report, the association with CIS-R score does not indicate either the direction of causality or whether a causal relationship exists. *(Table 8.2)*

ii. Small primary support group and a severe lack of social support

The second analysis looked at people with the least social support as defined both by having a small network of close friends and relatives and being categorised as having a severe lack of perceived support.

Again, living arrangements and CIS-R score showed significant associations with this combination of circumstances. People living with other members of their family were least likely to have the lowest levels of social support and the odds were highest for informants living in institutional households. There was also a strong association with CIS-R score but this was only seen for people with a score of 18 or more. *(Table 8.3)*

iii. Smoking, drinking and drug use

The third analysis was concerned with lifestyle behaviours - smoking, drinking and drug use. The first model tested looked at characteristics associated with heavy smoking, drinking more than the recommended sensible level and having used illegal drugs. This combination might indicate an increased risk to some aspects of health or a greater dependence on substances which affect one's mental state. No characteristics were significantly associated with this combination of measures, nor with just being a heavy smoker and heavier drinker. The negative results suggest that the people falling into these categories are a relatively varied group so that no strong associations emerge. This is perhaps to be expected given that the behaviours are prone to change over time and changes are particularly likely in response to health problems and advice from health professionals.

Table 8.1 Associations between variables describing the circumstances of people with psychosis

		On medication	Used any services	4 or more ADL difficulties	Unable to work	Small support group	Severe lack of support	Few leisure activities	Heavy smoker	Heavy drinker	Used drugs
On medication	No (ref)	/									
	Yes	/	***					(*)	***		
Used health services	No (ref)		/								
	Yes	***	/								
ADL difficulties	None (ref)			/							
	1			/	**						
	2 or 3		*	/	***			**	*		
	4 or more		*	/	***			***	*		
Economic activity	Working (ref)				/						
	Unemployed				/						
	Unable to work	*	**	***	/	**	**	**	**		
	Other inactive	*			/			*			
Primary support group	0–3 people				***	/	***	***			
	4–8 people					/	*	**			
	9 or more people (ref)					/					
Social support	Severe lack				**	**	/	*			
	Moderate lack						/				
	No lack (ref)						/				
No. of leisure activities	0–3			***	***	***	***	/			
	4–9	**		*	**	**	***	/			
	10 or more							/			
Cigarette smoking	Non-smoker (ref)								/		
	Light/moderate								/		
	Heavy smoker	**			***				/	*	*
Alcohol consumption	Abstainer/occasional (ref)									/	
	Light/moderate								**	/	
	Above sensible level								**	/	
Drug use	Non-user (ref)										/
	Used cannabis	(**)		**					*		/
	Used other drugs only										/
CIS–R score	0–11 (ref)										
	12–17	(***)									
	18 or more		**	***	**						**
Living arrangements	Private household (ref)										
	Supported accommodation	*			***		***		***		

(ref) reference category
Significance level of difference from reference category: *p<0.05 **p<0.01 ***p<0.001
(*) odds significantly less than those for reference category

Table 8.2 Characteristics associated with having 4 or more ADL difficulties and being classified as unable to work because of long-term illness

		Adjusted OR
Housing tenure	Owned	1.0
	Rented	2.2*
	Supported accommodation	2.5**
CIS–R score	0–11	1.0
	12–17	3.0**
	18 or more	5.2***

*p<0.05 **p<0.01 ***p<0.001

Table 8.3 Characteristics associated with having a small primary support group and a severe lack of perceived social support

		Adjusted OR
Family circumstances	Supported accommodation	3.4***
	Not with family	2.4*
	With family	1.0
CIS–R score	0–11	1.0
	12–17	1.0
	18 or more	2.3**

*p<0.05 **p<0.01 ***p<0.001

8.3 Case studies

So far this report has used multi-variate analysis to look at characteristics associated with different behaviours and this chapter has followed a similar pattern to look at inter-relationships between the main measures. This approach has some limitations both because of the need to define a separate model for the large number of possible combinations which might occur and because it does not give an overview of the range of circumstances experienced by people with psychosis. This can be done more effectively by using a qualitative approach and looking at case studies of individuals in the sample.

The rest of this chapter therefore presents a small number of case studies which concentrate on the measures used throughout the report. The individuals selected are not representative of the sample but are illustrative of the range of people and circumstances on which the earlier analyses have been based. Cases were selected to cover a range of responses on the main topics covered in this report and the criteria for selection are given at the beginning of each case study. Some biographical details have been changed to preserve the anonymity of the informants.

Notes and references

1 Thomas M., Goddard E., Hickman M. and Hunter P. (1994) *1992 General Household Survey*, HMSO: London.

Case Study Number 1

Basis for selection
On antipsychotic medication

Profile
A single man in his twenties living in supported accommodation.

On the CIS-R, he was assessed as having significant neurotic symptoms in 9 areas and a total score of more than 18.

He suffered from paranoid schizophrenia and was on antipsychotic medication. This included both oral medication and depot injections.

Within the previous year he had consulted his GP and visited an out-patients department. He also received regular visits from a community psychiatric nurse.

The subject reported difficulties with four activities of daily living - using transport, medical care, household activities and dealing with paperwork. He received help with the first two activities from the community psychiatric nurse and a social worker.

He had a primary support group of between four and eight people, felt a moderate lack of social support and was involved in more than three leisure activities.

The subject was permanently unable to work and received an Invalidity Pension and a Disability Working Allowance.

He was classified as a heavy smoker and light drinker, and had taken hypnotics and cannabis in the past year.

Case study number 2

Basis for selection
On antidepressant medication

Profile
A lone mother living in privately rented accommodation.

On the CIS-R, she had a total score of 18 or more and had significant neurotic symptoms in five areas.

The subject screened positive on the Psychosis Screening Questionnaire. She was on antidepressant medication and said that she sometimes took more than the stated dose of this medicine.

Within the previous year she had consulted her GP but had not had an inpatient stay or visited an out-patients department for a mental or nervous complaint. She was visited at home by a community psychiatric nurse.

The subject had no difficulty with any activities of daily living.

She had a large number of relatives and friends whom she regarded as close, felt no lack of social support and took part in a large range of leisure activities. At the time of interview, the subject was working part time and receiving Child Benefit, One-Parent Benefit and Family Credit.

She smoked heavily, drank more than the recommended sensible level and had not taken drugs.

Case Study Number 3

Basis for selection
Refused medication

Profile
A single man in his twenties who lived by himself. On the CIS-R he had a score of less than 12 with significant neurotic symptoms in 3 areas.

The subject had had a nervous breakdown and was on antipsychotic medication. He had previously stopped other treatment of his own accord and he had turned down another type of medication because he was worried about side effects and did not like the treatment.

In the past year he had seen his GP and had an in-patient stay for his mental health problems. He used to make regular out-patient visits to a day centre but stopped these of his own accord. He received occasional visits from a community psychiatric nurse but had turned down consultations with a psychiatrist because he did not think it could help. For the same reason, he sometimes did not seek help when he or others around him thought that he should.

He had no difficulty with activities of daily living.

The subject had more than three close relatives and friends but felt a severe lack of social support. He took part in between four and nine leisure activities. He was classed as permanently unable to work and was receiving Income Support.

The subject was a non-smoker, rarely drank alcohol and did not take drugs.

Case Study Number 4

Basis for selection
Used all main health services in past year: GP, in-patient stay, out-patient and domiciliary visits.

Profile
A middle-aged woman in her fifties living on her own.

On the CIS-R she had a score of less than 12 with significant neurotic symptoms in two areas.

She was prescribed both antipsychotic and antidepressant medication.

In the past year she had, on at least one occasion, consulted her GP, had an in-patient stay and made an out-patient visit for a mental, nervous or emotional problem. She also received visits from a community psychiatric nurse.

She reported no difficulties with daily living.

The subject said she felt close to between four and eight people and felt no lack of social support. She took part in a large number of leisure activities.

The subject worked full time and had done so for a number of years.

At the time of interview she did not smoke and had stopped drinking for health reasons.

Case Study Number 5

Basis for selection
Used no health services in past year for a mental, nervous or emotional problem.

Profile
A single man in his thirties living in temporary accommodation.

On the CIS-R, he was assessed as having significant neurotic symptoms in 12 of the 14 areas and had a score of more than 18.

He was prescribed antipsychotic medication but sometimes did not take the drugs because of the side effects . He had previously stopped taking other medication for the same reason.

In the past year he had not seen a GP, visited an out-patients department, made an in-patient stay or received a domiciliary visit. He avoided contact with doctors.

The subject said that he felt close to fewer than four people and had a severe lack of social support.

He had not worked for about ten years and assessed himself as permanently unable to work. He was on Income Support.

The subject was a moderate smoker and had stopped drinking for health reasons.

Case Study Number 6

Basis for selection
Refused services

Profile
A widowed woman living with her children.

On the CIS-R she had a score of more than 18 with significant neurotic symptoms in seven areas.

The subject reported that her GP had diagnosed clinical depression and was on antidepressant medication. She screened positive on the Psychosis Screening Questionnaire.

In the past year she had made regular out-patient visits to see a psychiatrist. She was offered other visits but had turned them down because she felt she did not need help. The subject was hesitant to seek professional advice for a mental health problem and reported feeling embarrassed to discuss her problem. She was afraid what family or friends might think and felt that she should be able to cope with her problem herself.

The subject reported no difficulties in relation to activities of daily living.

She felt close to fewer than four people and felt a moderate lack of social support. She participated in fewer than four leisure activities and felt that she had no social life.

At the time of interview, the subject was in full-time work and had had no time off sick in the previous year. She received Child Benefit and Widowed Mother's Allowance.

She was classified as a heavy smoker and a fairly heavy drinker. She had never taken illegal or extra-medical drugs.

Case Study Number 7

Basis for selection
Difficulties with all activities of daily living

Profile
A single man living with his family.

On the CIS-R he had a score of more than 18 with significant neurotic symptoms in nine of the 14 areas.

The subject said he suffered from two longstanding physical complaints as well as schizophrenia. He was on seven types of medication including antipsychotic drugs and depot injections, and reported that he sometimes forgot to take the antipsychotic drugs.

During the previous year he had regularly consulted his GP and made out-patient visits, and had also had an in-patient stay for his mental health problem. He was also visited by a community psychiatric nurse.

He had difficulty and needed help with all activities of daily living. His family were his main source of help but a voluntary worker helped with getting about and using transport.

The subject said that he was close to fewer than four people and felt a severe lack of social support. He took part in between four and nine leisure activities.

The subject was permanently unable to work and received Invalidity Benefit and Disability Living allowance.

He was a non-smoker and occasional drinker.

Case study number 8

Basis for selection
No ADL difficulties

Profile
A single man in his fifties living alone.

On the CIS-R he had a score of less than 12 with neurotic symptoms in only one area. He said he suffered from manic depression and was on antipsychotic medication.

The subject had seen his GP in the past year for his mental condition but had not had an in-patient stay, out-patient or domiciliary visits.

He had no difficulty with any activity of daily living and participated in a wide range of leisure activities (10 or more). He felt close to a large number of people and felt no lack of social support.

The subject was working full time.

He was classified as a light smoker and had given up drinking.

Case Study Number 9

Basis for selection
Good social functioning

Profile
A middle-aged married woman living with her husband and children.

On the CIS-R she was assessed as having no neurotic symptoms.

The subject said that she suffered from severe depression and she was on both antipsychotic and antidepressant medication.

Her only contact with health services in the past year had been with her GP.

The subject reported no difficulties with activities of daily living.

She had a very active social life and took part in 10 or more of the listed activities either in or out of the home. The subject felt close to a number of friends and relatives and had no lack of social support.

She had previously had a job but, at the time of interview, was looking after her home and family.

The subject was a non-smoker and classified as a moderate drinker.

Case Study Number 10

Basis for selection
Poor social functioning.

Profile
A young single woman living alone.

On the CIS-R she had a score of more than 18 with significant neurotic symptoms in eight areas.

The subject had been given a diagnosis of manic depression. She was on antipsychotic and antidepressant medication. She reported that she sometimes stopped taking her medication when she felt she did not need it. She had also sometimes taken more than the stated dose.

In the previous year she had been an in-patient and seen a psychiatrist and a social worker regularly.

The subject reported difficulties with two activities of daily living - dealing with paperwork and managing money. The social worker helped her with budgeting and paying bills.

She felt close to fewer than four relatives and friends and felt a severe lack of social support. She took part in between four and nine leisure activities and also went to a Day Centre, a club for people with mental health problems, and a social club.

At the time of interview, the subject was unemployed but looking for work, preferably a sheltered job or part time work. She received Income Support.

She smoked heavily and was classified as a moderate drinker. She had not used illegal drugs.

Case Study Number 11

Basis for selection
Working full time

Profile
A divorced woman living with her children.

On the CIS-R she had a score just above the threshold level of 12 with significant neurotic symptoms in three areas.

She said that she had psychotic depression which occurred every four to five years. She was on no medicinal regime and said that she did not like to take drugs because of their side effects.

The subject had seen her GP and had an in-patient stay for a mental health problem in the past year. She also received regular visits from a community psychiatric nurse. She said she was hesitant to seek professional advice for a mental health problem because she thought the problem would get better by itself.

The subject reported difficulties with practical activities and dealing with paperwork. A friend helped with practical problems and she would have liked someone to help her deal with letters and bills.

The subject said she felt close to between four and nine relatives and friends and felt no lack of social support. She said she participated in 17 different leisure activities (out of a maximum of 26) in and outside her home. She also attended a club for people with mental health problems and an Adult Education centre.

At the time of interview, the subject was in full-time work but had been off work for more than half of the previous year due to ill health. She received Child Benefit and One-Parent Benefit.

She was a non-smoker, occasional drinker and did not take drugs.

Case Study Number 12

Basis for selection
Working part time

Profile
A woman in her sixties living with her husband.

On the CIS-R she had a score of less than 12 with significant neurotic symptoms in 2 areas.

She said she suffered from schizophrenia and was on antipsychotic medication. She reported that she sometimes took more than the stated dose of one of the drugs because she felt she needed to control her symptoms. She had received depot injections but had refused to continue with this form of treatment because she did not like it.

In the past year she had not seen her GP or been an in-patient, but had made out-patient visits. She received no domiciliary visits.

She reported having difficulties dealing with paperwork and was helped by her husband.

The subject felt no lack of social support and had more than three close friends. She took part in between four and nine leisure activities.

At the time of interview she was self employed and worked part time. She was classified as a light smoker and a light drinker.

Case Study Number 13

Basis for selection
Unemployed

Profile
A single woman living with her mother.

On the CIS-R she had a score of more than 18 with significant neurotic symptoms in six areas.

The subject reported that she had an anxiety problem and that her doctor said she suffered from depression. She was on antipsychotic and antidepressant medication.

In the previous year she had seen her GP for a mental health problem and regularly made visits to an out- patients department. She said she sometimes avoided seeking help when she needed it because she did not think that anyone could help.

In relation to activities of daily living, the subject reported difficulties with getting out and about and using transport but she was helped by her family.

The subject said she felt close to between four and eight people and had no lack of social support. She participated in more than four leisure activities.

At the time of interview, the subject was unemployed and seeking work. She was in receipt of Unemployment Benefit.

She was a non-smoker and a light drinker.

Case Study Number 14

Basis for selection
Permanently unable to work

Profile
A middle-aged married woman living with her husband and children.

On the CIS-R she had a score of more than 18 with significant neurotic symptoms in 10 areas.

The subject said she suffered from depression and she screened positive on the Psychosis Screening Questionnaire. She was not prescribed either antipsychotic or antidepressant medication.

In the past 12 months, she had visited her GP for a mental health problem but had made no in-patient stays, out-patient or domiciliary visits. She said that she sometimes did not seek professional help when she needed it because she did not want to be put back on antidepressants.

She had difficulties with four activities of daily living - getting out and about, household activities, dealing with paperwork and managing money. Her family helped her with all these tasks.

The subject felt that she had no lack of social support and had several close relatives and friends. She took part in between four and nine leisure activities.

She regarded herself as permanently unable to work and she received Income Support.

The subject was classified as a heavy smoker and drank more than the recommended sensible level. She had not taken drugs in the past year.

Case Study Number 15

Basis for selection
Heavy smoker and drinker and drug user

Profile
A young man living with other unrelated adults.

On the CIS-R he had a score of 22 with significant neurotic symptoms in eight areas. He reported that a specialist had said that he had paranoia. He was on no oral medication or injections.

The subject had seen his GP for a mental health problem and had had just one out-patient appointment in the last year. He avoided seeing a doctor or seeking other professional help when he felt he needed to because he had a fear of public places. He also believed that there was nothing doctors could do for him.

The subject had difficulty with three activities of daily living: getting out and about, medical care and managing money. He said he did not need help with getting out or with medical matters but someone helped him financially.

He felt no lack of social support and had between four and eight close friends. He took part in between four and nine leisure pursuits.

The subject had a part-time job.

He was a heavy smoker, drank more than the recommended sensible level and was classed as alcohol dependent. In the past year he had used hypnotics, cannabis and other drugs.

Case Study Number 16

Basis for selection
Non-smoker, non-drinker and not a drug user

Profile
An older married woman living with her family.

On the CIS-R she had a score below the threshold level of 12 with significant neurotic symptoms in four areas.

The subject said that she had bad nerves and felt depressed. She was taking antidepressant medication and was on antipsychotic depot injections. She said that she sometimes forgets to take the antidepressant tablets and on other occasions would take more than her normal dose.

In the past year she had not seen her GP for a mental or emotional problem, and had made no in-patient stays or out-patient visits. She was regularly visited at home by a community psychiatric nurse.

The subject reported difficulties with most of the activities of daily living. She said she was helped by her family.

She felt close to several relatives and friends and had no lack of social support. However she took part in fewer than four leisure activities.

At the time of interview, the subject stayed at home to look after family.

She had not worked for twenty years. She had never smoked, did not drink and had never taken drugs.

Appendix A Measuring psychiatric morbidity

A1 Identifying neurotic psychopathology

To obtain the prevalence of both symptoms and diagnoses of diagnoses of neurotic psychopathology, the revised version of the Clinical Interview Schedule (CIS–R) was chosen.[1] The CIS–R is made up of 14 sections, each section covering a particular area of neurotic symptoms.

Each section within the interview schedule starts with a variable number of mandatory questions which can be regarded as sift or filter questions. They establish the existence of a particular neurotic symptom in the past month. A positive response to these questions leads the interviewer on to further enquiry giving a more detailed assessment of the symptom in the past week. The symptom is assessed in terms of frequency, duration, severity and time since onset. The informant's responses to these questions determine the score on each section. More frequent and more severe symptoms result in higher scores.

The minimum score on each section is 0, where the symptom was either not present in the past week or was present only in mild degree. The maximum score on each section is 4 (except for the section on Depressive ideas which has a maximum score of 5).

- Summed scores from all 14 sections range between 0 and 57.
- The overall threshold score for significant psychiatric morbidity is 12.
- Symptoms are regarded as significant if they have a score of 2 or more.

The elements contributing to scores on each symptom are shown below. Any combination of the elements produce the section score.

Fatigue

Scores relate to fatigue or feeling tired or lacking in energy in the past week.

Score one for each of:

- Symptom present on four days or more
- Symptom present for more than three hours in total on any day
- Subject had to push him/herself to get things done on at least one occasion
- Symptom present when subject doing things he/she enjoys or used to enjoy at least once

Sleep problems

Scores relate to problems with getting to sleep, or otherwise, with sleeping more than is usual for the subject in the past week.

Score one for each of:

- Had problems with sleep for four nights or more
- Spent at least 4 hours trying to get to sleep on the night with least sleep
- Spent at least 1 hour trying to get to sleep on the night with least sleep
- Spent three hours or more trying to get to sleep on four nights or more
- Slept for at least 4 hours longer than usual for subject on any night
- Slept for at least 1 hour longer than usual for subject on any night
- Slept for more than three hours longer than usual for subject on four nights or more

Irritability

Scores relate to feelings of irritability, being short-tempered or angry in the past week.

Score one for each of:

- Symptom present for four days or more
- Symptom present for more than one hour on any day

- Wanted to shout at someone (even if subject had not actually shouted)
- Had arguments, rows or quarrels or lost temper with someone and felt it was unjustified on at least one occasion

Worry

Scores relate to subject's experience of worry in the past week, other than worry about physical health.

Score one for each of:

- Symptom present on four or more days
- Has been worrying too much in view of circumstances
- Symptom has been very unpleasant
- Symptom lasted over three hours in total on any day

Depression

Applies to subjects who felt sad, miserable or depressed or unable to enjoy or take an interest in things as much as usual, in the past week. Scores relate to the subject's experience in the past week.

Score one for each of:

- Unable to enjoy or take an interest in things as much as usual
- Symptom present on four days or more
- Symptom lasted for more than three hours in total on any day
- When sad, miserable or depressed subject did not become happier when something nice happened, or when in company

Depressive ideas

Applies to subjects who had a score of 1 for depression. Scores relate to experience in the past week.

Score one for each of:

- Felt guilty or blamed him/herself at least once when things went wrong when it had not been his/her fault
- Felt not as good as other people
- Felt hopeless
- Felt that life isn't worth living
- Thought of killing him/herself

Anxiety

Scores relate to feeling generally anxious, nervous or tense in the past week. These feelings were not the result of a phobia.

Score one for each of:

- Symptom present on four or more days
- Symptom had been very unpleasant
- When anxious, nervous or tense, had one or more of following symptoms:
  heart racing or pounding
  hands sweating or shaking
  feeling dizzy
  difficulty getting breath
  butterflies in stomach
  dry mouth
  nausea or feeling as though he/she wanted to vomit
- Symptom present for more than three hours in total on any one day

Obsessions

Scores relate to the subject's experience of having repetitive unpleasant thoughts or ideas in the past week.

Score one for each of:

- Symptom present on four or more days
- Tried to stop thinking any of these thoughts
- Became upset or annoyed when had these thoughts

- Longest episode of the symptom was ¼ hour or longer

Concentration and forgetfulness

Scores relate to the subject's experience of concentration problems and forgetfulness in the past week.

Score one for each of:

- Symptoms present for four days or more
- Could not always concentrate on a TV programme, read a newspaper article or talk to someone without mind wandering
- Problems with concentration stopped subject from getting on with things he/she used to do or would have liked to do
- Forgot something important

Somatic symptoms

Scores relate to the subject's experience in the past week of any ache, pain or discomfort which was brought on or made worse by feeling low, anxious or stressed.

Score one for each of:

- Symptom present for four days or more
- Symptom lasted more than three hours on any day
- Symptom had been very unpleasant
- Symptom bothered subject when doing something interesting

Compulsions

Scores relate to the subject's experience of doing things over again when subject had already done them in the past week.

Score one for each of:

- Symptom present on four days or more
- Subject tried to stop repeating behaviour
- Symptom made subject upset or annoyed with him/herself
- Repeated behaviour three or more times when it had already been done

Phobias

Scores relate to subject's experience of phobias or avoidance in the past week

Score one for each of:

- Felt nervous/anxious about a situation or thing four or more times
- On occasions when felt anxious, nervous or tense, had one or more of following symptoms:
  heart racing or pounding
  hands sweating or shaking
  feeling dizzy
  difficulty getting breath
  butterflies in stomach
  dry mouth
  nausea or feeling as though he/she wanted to vomit
- Avoided situation or thing at least once because it would have made subject anxious, nervous or tense
- Avoided situation or thing four times or more because it would have made subject anxious, nervous or tense

Worry about physical health

Scores relate to experience of the symptom in the past week.

Score one for each of:

- Symptom present on four days or more
- Subject felt he/she had been worrying too much in view of actual health
- Symptom had been very unpleasant
- Subject could not be distracted by doing something else

Panic

Applies to subjects who felt anxious, nervous or

tense in the past week and the scores relate to the resultant feelings of panic, or of collapsing and losing control in the past week.

Score one for each of:

- Symptom experienced once
- Symptom experienced more than once
- Symptom had been very unpleasant or unbearable
- An episode lasted longer than 10 minutes

A2 Identifying psychotic psychopathology

The criteria used for identifying people with psychosis are described in Section 1.4 of this report.

List of conditions: see Reference card A

List of relevant drugs: see Reference card B

Psychosis Screening Questionnaire: see pages 66 to 68

Notes and references

1 Lewis, G. and Pelosi, A. J., *Manual of the Revised Clinical Interview Schedule, (CIS–R)*, June 1990, Institute of Psychiatry.

Lewis, G., Pelosi, A. J., Araya, R. C. and Dunn, G., (1992) Measuring Psychiatric disorder in the community: a standardized assessment for use by lay interviewers, *Psychological Medicine*, 222, 465–486

Reference card A

Auditory hallucinations
Bipolar affective disorder
Catatonic schizophrenia
Chronic schizophrenia
Hallucinations
Hearing voices
Hebephrenic schizophrenia
Hypomania
Mania
Manic depression
Manic depressive psychosis
Mental illness
Mentally disturbed
Mild psychosis
Mild schizophrenia
Mood swings
Neuroleptic
Paranoia
Paranoid schizophrenia
Psychosis
Psychotic related disorder
Psychotic tendencies
Schizo-affective disorder
Schizophrenia
Schizophrenic affective disorder
Simple schizophrenia
Voices

Reference card B

Anquil
Benperidol
Camcolit
Chlorpromazine
Clopixol acuphase
Clopixol
Clozapine
Clozaril
Depixol
Dolmatil
Dozic
Droleptan
Droperidol
Fentazin
Fluanxol
Flupenthixol
Flupenthixol decanoate
Fluphenazine hydrochloride
Fluphenazine decanoate
Fluphenazine enanthate
Fluspirilene
Fortunan
Haldol
Haldol decanoate
Halperidol
Largactil
Liskonum
Litarex
Lithium
Loxapac
Loxapine
Melleril
Methotrimeprazine
Modecate
Moditen
Moditen ethanate
Neulactil
Nozinan
Orap
Oxypertine
Pericyazine
Perphenazine
Phasal
Pimozide
Piportil
Pipothiazine palmitate
Priadel
Prochloperazine
Promazine hydrochloride
Redeptin
Remoxipride
Roxiam
Serenace
Sparine
Stelazine
Sulphiride
Sulpitil
Thioridazine
Trifluoperazine
Trifluperidol
Zuclopenthixol dihydrochloride
Zuclopenthixol acetate
Zuclopenthixol decanoate

Antipsychotic drugs
Antipsychotic injections
Depot injections
Antimanic drugs

Psychosis Screening Questionnaire

[*] **P. PSQ**

P1. **Over the past year,** have there been times when you felt very happy indeed without a break for days on end?

Yes	1	→ (a)
Unsure	2	→ P2
No	3	→ P2

(a) Was there an obvious reason for this?

Yes	1	→ P2
Unsure	2	→ P2
No	3	→ (b)

(b) Did your relatives or friends think it was strange or complain about it?

Yes	1	→ **Screen Positive, Go to P6**
Unsure	2	→ P2
No	3	→ P2

P2. **Over the past year**, have you ever felt that your thoughts were directly interfered with or controlled by some outside force or person?

Yes	1	→ (a)
Unsure	2	→ P3
No	3	→ P3

(a) Did this come about in a way that many people would find hard to believe, for instance, through telepathy?

Yes	1	→ **Screen Positive, Go to P6**
Unsure	2	→ P3
No	3	→ P3

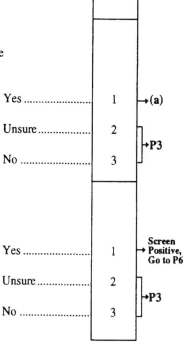

P3. **Over the past year,** have there been times when you felt that people were against you?

Yes	1	→(a)
Unsure	2	→P4
No	3	→P4

(a) Have there been times when you felt that people were deliberately acting to harm you or your interests?

Yes	1	→(b)
Unsure	2	→P4
No	3	→P4

(b) Have there been times you felt that a group of people was plotting to cause you serious harm or injury?

Yes	1	→ Screen Positive, Go to P6
Unsure	2	→P4
No	3	→P4

P4 **Over the past year,** have there been times when you felt that something strange was going on?

Yes	1	→(a)
Unsure	2	→P5
No	3	→P5

(a) Did you feel it was so strange that other people would find it very hard to believe?

Yes	1	→ Screen Positive, Go to P6
Unsure	2	→P5
No	3	→P5

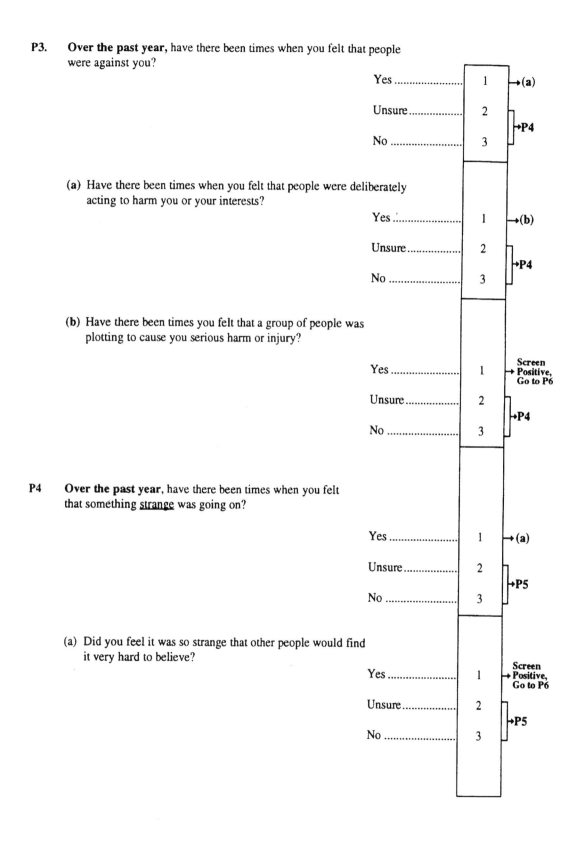

P5. **Over the past year,** have there been times when you heard or saw things that other people couldn't?

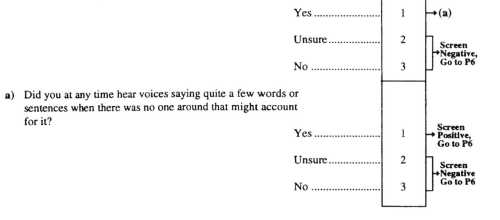

Yes	1	→ **(a)**
Unsure	2	→ **Screen Negative, Go to P6**
No	3	

a) Did you at any time hear voices saying quite a few words or sentences when there was no one around that might account for it?

Yes	1	→ **Screen Positive, Go to P6**
Unsure	2	→ **Screen Negative Go to P6**
No	3	

P6. **Interviewer check**

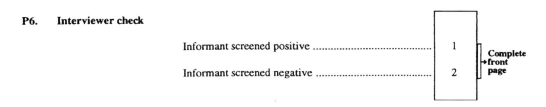

Informant screened positive	1	→ **Complete front page**
Informant screened negative	2	

Appendix B Multiple logistic regression and odds ratios

Logistic regression analysis

This report uses logistic regression analysis to test for associations between a dependent dichotomous variable, such as having consulted a GP or not, and a number of categorical independent variables, such as age group or CIS-R score. A forward stepwise method of logistic regression was used in which the variable which is most strongly associated with the dependent variable is selected into the model first, and then the test is repeated until all variables which have a significant effect are included in the model. Multiple logistic regression estimates the effect of any independent variable after controlling for the confounding effects of all other variables in the model.

The models reported here are main effects models and no testing was carried out to identify significant interactions. The tables show only those variables which were significantly associated with the dependent measure.

Interpretation of odds ratios and significance levels

Logistic regression produces an estimate of the odds of an event occurring when an individual is in a particular category of an independent variable compared to a reference category of that variable. The odds are defined as the ratio of the probability of the event occurring compared with the probability of it not occurring. If the probability of an event is p, the odds are p/(1-p). The factor by which the odds of an event differ for people in a particular category compared with those in the reference category (OR of 1.0) is shown in tables by the Adjusted Odds Radio (OR). The Adjusted OR controls for the effects of other variables in the statistical model, eg. age or occupation type.

The tables indicate which odds ratios are significantly different ($p<0.05$) from 1.00 using conventional markings for the significance level. Thus results marked with an asterisk would normally be considered to be unlikely to have occurred simply by chance but rather to indicate a real difference between groups. However, because the models reported here have been run on unweighted data (see Chapter 1), the odds ratios cannot be used as estimates for the population of all psychotic people, nor are the significance tests strictly valid. Nonetheless the significance levels are shown in the tables because they give an indication of the reliability of the results. Thus, results marked ** are considered to be more reliable than those marked *.

Glossary of survey definitions and terms

Adults
In this survey adults were defined as persons aged 16 to 64 years.

Antipsychotic drugs
These are also known as ''neuroleptics''. In the short term they are used to quieten disturbed patients whatever the underlying psychopathology. See Depot Injections

Depot injections
When antipsychotic medication is given by injections on a weekly or monthly basis, these are sometimes termed depot injections.

Economic activity status
Four categories were identified for the analysis in this report: working; unemployed; unable to work because of long-term sickness of disability; other economically inactive.

Working adults - having done paid work in the seven days ending the Sunday before the interview, either as an employee or self-employed, including those who were not actually at work but had a job they were away from. Anyone on a Government scheme which was employer based was also classed as working last week.

Unemployed - waiting to take up a job that had already been obtained, looking for work, and people who intended to look for work but were prevented by temporary ill-health, sickness or injury.

Permanently unable to work - people unable to work due to health or emotional problems or disablement. Applied only to those under state retirement age, ie to men aged 16 to 64 and to women aged 16 to 59.

Other economically inactive
This category comprised three main categories of people:

- Going to school or college: This only applied to people who were under 50 years of age.

- Retired: used only if stopped work at the age of 50 or over.

- Looking after the home or family: anyone who was mainly involved in domestic duties, provided this person had not already been coded in an earlier category.

Educational level
Educational level was based on the highest educational qualification obtained and was grouped as follows:

i. A levels or above
Degree (or degree level qualification)
Teaching qualification
HNC/HND, BEC/TEC Higher, BTEC Higher
City and Guilds Full Technological Certificate
Nursing qualifications: (SRN, SCM, RGN, RM, RHV, Midwife)
A-levels/SCE higher
ONC/OND/BEC/TEC/not higher
City and Guilds Advanced/Final level

ii. Other qualifications
GCE O-level (grades A-E)
GCSE (grades A-G)
CSE (all grades)
SCE Ordinary (bands A-E)
Standard grade (levels 1-5)
SLC Lower SUPE Lower or Ordinary
School certificate or Matric
City and Guilds Craft/Ordinary level
Clerical or commercial qualifications
Apprenticeship
Other qualifications

iii. No qualifications

Ethnic Group
Household members were classified into nine groups by the person answering Schedule A:

White: Black - Caribbean; Black - African; Black - Other; Indian; Pakistani; Bangladeshi; Chinese; None of these

For this report the eight groups other than ''White'' were combined.

Family circumstances
In order to classify the relationships of the subject to

other members of the household, the household members were divided into family units. For this report, informants living in private households were classified as either belonging to a family or not. In this case a family was defined as:
(a) a married or cohabiting couple living either on their own or with the never-married children of one or other partner;
(b) a lone parent with his or her never-married children.

People who did not belong to one of these categories were classified as not living with a family. The category largely comprises people who lived alone and those who lived with other non-related adults.

Household
The standard definition used in most surveys carried out by OPCS Social Survey Division, and comparable with the 1991 Census definition of a household, was used in this survey. A household is defined as a single person or group of people who have the accommodation as their only or main residence and who either share one meal a day or share the living accommodation. (See E McCrossan *A Handbook for interviewers*. HMSO: London 1985.)

Living arrangements
The analysis in this report covers individuals who were defined as living in households as opposed to large institutions such as hospitals. The sample covers two main types of accommodation.

Supported accommodation: residential accommodation which was attached to an institution catering primarily for those with mental health problems. This includes group homes, recognised lodgings and supervised ordinary housing.

Private households: people living independently in residental households which are not supervised or attached to an institution.

Marital status
Informants were categorised according to their own perception of marital status. Married and cohabiting took priority over other categories. Cohabiting included anyone living together with their partner as a couple.

Occupation type
Based on the Registrar General's *Standard Occupational Classification*. Volume 3 OPCS London: HMSO 1991, social class was first ascribed on the basis of the following priorities.

a) Social class was based on the informant's own occupation unless the informant was a married or cohabiting woman whose partner was in the same household. In such cases, the spouse or partner's occupation was used unless he had never worked, in which case the woman's own occupation was used.
b) Social class was based on the informant's (or spouse's) current occupation or, if the informant (or spouse) was unemployed or economically inactive at the time of interview but had previously worked, it was based on the most recent occupation.
c) Social class was not determined where the informant (and spouse) had never worked, or was a full-time student or their occupation was inadequately described.

The classification of occupation type used in the tables is as follows:

Descriptive definition	*Social class*
Non-manual	I, II or III non-manual
Manual	III manual, IV or V or not known

Physical illness
Informants were asked 'Do you have any longstanding illness, disability or infirmity? By longstanding I mean anything that has troubled you over a period of time or that is likely to affect you over a period of time?'

Those that answered yes to this question were then asked 'What is the matter with you?'; interviewers were asked to try and obtain a medical diagnosis, or to establish the main symptoms. From these responses, illnesses were coded to the site or system of the body that was affected, using a classification system that roughly corresponded to the chapter headings of the International Classification of Diseases (ICD-10). Some of the illnesses identified were mental illnesses and these were excluded from the classification of physical illness. Physical illness did, however include physical disabilities and sensory complaints such as eyesight and hearing problems.

Psychiatric morbidity
The expression psychiatric morbidity refers to the degree or extent of the prevalence of mental health problems within a defined area.

Tenure

Two tenure categories for private households were used in this report

Owner occupied - buying with a mortgage or loan or owned outright. This includes co-ownership and shared ownership schemes.

Rented - rented from a local authority, New Town corporations, housing association or other organisation or from an individual.

Printed in the United Kingdom for HMSO
Dd. 0302532 C18 5/96 48186